SPECIAL PAPERS IN PALAEONTOLOGY NO. 85

THE PHYLOGENY OF POST-PALAEOZOIC ASTEROIDEA (NEOASTEROIDEA, ECHINODERMATA)

BY

ANDREW SCOTT GALE

with 27 plates, 36 text-figures and 8 tables

THE PALAEONTOLOGICAL ASSOCIATION
LONDON

April 2011

CONTENTS

[Special Papers in Palaeontology, 85, 2011, pp. 5–112]

Abstract: The skeletal morphology and homologies of 24 extant asteroid taxa and one Palaeozoic outgroup (*Calliasterella mira*; Upper Carboniferous, Moscow area, Russia) are described in detail. This information is used to construct a character matrix to identify deep relationships within the Neoasteroidea. Particular attention is paid to the morphology and homologies of the character-rich ossicles of the mouth frame (orals, circumorals, odontophore) and of the ambulacral groove (adambulacrals, ambulacrals). In addition, new homologies between alveolar and forcipulate pedicellariae are identified. Soft-tissue characters are reviewed from the literature. Simultaneous unconstrained cladistic analysis of 128 characters yielded six equally parsimonious trees, and a consensus tree of these is presented. This places the Paxillosida as the most basal neoasteroid group, sister taxon to the Surculifera, which include all other neoasteroids. The Spinulosida (used here in its original sense) are a well-supported monophyletic group, and a newly identified clade here named the Tripedicellaria, unified by characters of the pedicellariae, includes both the Valvatida and Forcipulatida. 'Valvatids' are a paraphyletic group of taxa that are sister group to the Forcipulatida. The fossil record of the asteroid families to which the 24 investigated species belong is reviewed, and the stratigraphical information is used to constrain the positions of the nodes on the consensus tree. The enigmatic asteroid fossil record from the Triassic is reviewed, and it is concluded that the order Trichasteropsida is a poorly understood assemblage of diverse morphologies of which the relationships with other taxa are uncertain. Middle Triassic (Muschelkalk) asteroids are identified as neoasteroids *incertae sedis* and assigned to two families: Trichasteropsidae and Migmasteridae fam. nov. Well-preserved asteroid ossicles from the Carnian of the Italian Dolomites are described and demonstrate that much of the neoasteroid radiation had taken place by the early Carnian (Late Triassic) because ossicles of taxa close to ophidiasterids and asterinids are present. It is concluded that the sudden appearance of diverse taxa with close affinities to modern families in the Early and Mid-Jurassic is an artefact of the extensive record of marine sediments of these ages and not of contemporaneous radiation of the neoasteroids. Detailed ossicular morphology is extremely useful and underused, both in the identification of asteroid relationships and for the assignation of fossil neoasteroids to extant families. The adaptive evolution of the neoasteroid skeleton is reviewed, and the neoasteroids are considered to be primitively infaunal and only secondarily adapted to epifaunal modes of life, the opposite of the case of post-Palaeozoic echinoids. It is concluded that *Xyloplax* is a true neoasteroid and most closely related to the Caymanostellidae. A new genus of Xyloplacidae, *Ankyloplax*, is erected. *Calliasterella mira*, from the Upper Carboniferous of Moscow, is redescribed. The first Jurassic goniopectinid (*Chrispaulia jurassica* sp. nov.), a benthopectinid (*Jurapecten hessi* gen. et sp. nov.) and two pterasterids (*Savignaster wardi* gen. et sp. nov., *Savignaster trimbachensis* gen. et sp. nov.) are described from the upper Oxfordian (Upper Jurassic) of the French Jura. *Terminaster cancriformis* is redescribed on the basis of new material and assigned to a new family, the Terminasteridae.

Key words: Asteroidea, neoasteroids, phylogeny, classification.

THE phylogeny of the class Asteroidea de Blainville, 1830 (Ordovician to the present day) has proved to be a contentious issue with little consensus on relationships, reflected by the existence of disparate classification schemes. Neontologists have traditionally used various modifications of Perrier's (1884, 1894) classification based largely upon the presence, absence and detailed morphology of the pedicellariae present in living forms (Paxillosida, Spinulosida, Forcipulatida and Valvatida), mixed variably with Sladen's (1889) two-part division of the class based on the size and development of the marginal ossicles (Phanerozonida and Cryptozonida). Spencer and Wright (1966) attempted to integrate an ordinal classification developed for extant asteroids with Spencer's scheme for Palaeozoic taxa, thus creating a comprehensive classification in which several orders extended from an Early Palaeozoic (Ordovician) radiation to the present day. Gale (1987) and Blake (1987) independently argued for the monophyly of post-Palaeozoic asteroids because they share a large number of synapomorphies, in particular of ossicles of the ambulacral groove.

Gale (1987) named this group the Neoasteroidea and suggested that it evolved from a single common ancestor, which crossed the Permo–Triassic boundary. In contrast, Blake (1987) included the Carboniferous asteroid *Calliasterella* in the otherwise exclusively post-Palaeozoic superorder Forcipulatacea and suggested that this group had diverged by the Carboniferous. The cladograms generated for post-Palaeozoic asteroids by Gale (1987) and Blake (1987) differ on major respects. Gale, in keeping with

by ANDREW SCOTT GALE

School of Earth and Environmental Sciences, University of Portsmouth, Burnaby Building, Burnaby Road, Portsmouth PO1 3QL, UK; e-mail a.gale@port.ac.uk tradition, considered the Paxillosida to be the basal neo-asteroid group and the Forcipulatida to be the most derived and used this presumption to determine polarity within the post-Palaeozoic taxa. Blake (1987, 1988b) argued that the Paxillosida in fact were highly specialized, and their apparent plesiomorphy in lacking suckered tube feet, brachiolarian larvae and the ability to feed by stomach eversion were secondary losses enabling them to live successfully on soft substrata.

Blake (1987) based the polarity of his cladogram upon the supposition that the Carboniferous *Calliasterella* was a member of the extant superorder Forcipulatacea, following the interpretation of Downey (1970) that *Calliasterella mira* fell close to the extant forcipulatid family Zoroasteridae Sladen, 1889. Phylogenies, developed in subsequent papers by Blake and co-workers, show broadly similar relationships (Blake *et al.* 2000; Blake and Hagdorn 2003), and the classification of Blake (1987) has been subsequently used by most neontologists (e.g. Clark and Downey 1992). Molecular studies have provided somewhat ambiguous and conflicting evidence of asteroid relationships, and disparate phylogenies do not show any consensual patterns of relationship (Knott and Wray 2000; Matsubara *et al.* 2004). However, two studies (Lafay *et al.* 1995; Wada *et al.* 1996) support the basal position of the paxillosids, but identify the group as paraphyletic. Matsubara *et al.* (2005) found support for the monophyly of paxillosids from amino acid and nucleotide sequences, but argued from the same evidence that the clade was not basal to the neoasteroids.

The asteroid internal skeleton is a complex structure composed of up to 20 ossicle and 10 spine types in any individual species (Turner and Dearborn 1972). These can be divided into serial ossicles, which increase in number during growth by addition, and nonserial ossicles, which are fixed in number at metamorphosis. Mooi and David (2000) further subdivided the serial ossicles into axial components, which are added at the interface with the terminal ossicle (ambulacral, adambulacral) and extraxial ossicles, added elsewhere (marginal, actinal, abactinal). The mouth frame and ambulacral column, in particular, have numerous muscle attachment sites and articular surfaces that provide potential homologies with which to test relationships between taxa, in a parallel fashion to the use of bones in studies of the vertebrate skeleton. However, rather little use has been made of these characters in the description of extant asteroids or in the study of phylogenetic relationships, with the notable exception of papers by Blake (1973) and Villier *et al.* (2004b). Well-preserved fossil asteroids commonly have skeletal elements fused by diagenesis, and internal features of ambulacral groove and mouth frame ossicles are seldom visible. Descriptions of fossil taxa have therefore concentrated on external morphological characters.

This paper presents the results of a new investigation into asteroid skeletal morphology, based upon partial preparation of specimens and complete dissociation of ossicles using sodium hypochlorite. Partial skeletons and individual ossicles were examined using both light microscope and SEM. Twenty-four extant species, each belonging to a separate family, were studied in detail for cladistic analysis (Table 1). These species encompass a large portion of the morphological range of post-Palaeozoic asteroids. The family-, genus- and species-level nomenclature follows Clark (1989, 1993, 1996) and Clark and Mah (2001). The Late Carboniferous *Calliasterella mira* was selected as an outgroup, because this is one of the few Palaeozoic asteroids for which almost all ossicle types can be examined in a dissociated state and thus can be compared individually with those of extant taxa.

Institutional abbreviations. Material described and illustrated, or referred to, in the text is housed in the following collections: BAS, British Antarctic Survey, Cambridge; MHI, Muschelkalk Museum, Ingelfingen; NHM, Natural History Museum, London, UK; NHMB; Naturhistorisches Museum Basel; PIN, Palaeontological Institute, Moscow; SMU, Southern Methodist University, Dallas, Texas; USNM, United States National Museum, Washington DC; UWA, Museum of the University of Western Australia, Perth.

HOMOLOGIES OF THE ASTEROID SKELETON

Terminology

Good descriptions, illustrations and discussions of terminologies for asteroid ossicles are to be found in Turner and Dearborn (1972) and Blake (1973). However, these two papers use substantially different nomenclatures and only describe in any detail asteroids belonging to three families, Luidiidae Sladen, 1889, Astropectinidae Gray, 1840 and Ctenodiscidae Sladen, 1889. The present study has developed a new nomenclature (Table 2) in which the two opposing surfaces making up articulations and muscle insertion sites are given the same name, based loosely upon the scheme used by Blake (1973). The equivalence of the terms used by Turner and Dearborn (1972), Blake (1973) and the present study are given in Tables 3–5.

The basic construction and terminology of the asteroid skeleton is shown in Text-figure 1 and Plate 1. The asteroid comprises discrete abactinal (= aboral) and actinal (= oral) surfaces; the abactinal skeleton is made up of abactinal ossicles, and the ambitus is defined in many taxa by paired, opposing marginal ossicles (infero- and superomarginals). Primary radial ossicles (centrale, primary radials and primary interradials) form a central circlet in the juvenile asteroid and may be visible in the

TABLE 1. Taxa used in this study, their locations and museum registration numbers.

Family	Genus, species	Locality	Collection
Calliasterellidae	*Calliasterella mira* (Trautschold)	U. Carboniferous, Moscow Basin	PIN, Moscow unregistered
Astropectinidae	*Astropecten irregularis* (Pennant)	NE Ireland	NHM EE 13560
Luidiidae	*Luidia ciliaris* (Philippi)	NE Ireland	NHM EE 13561
Radiasteridae	*Radiaster tizardi* (Sladen)	NE Atlantic, w. of Scotland	RSM Edinburgh NMSZ:1996.003
Ctenodiscidae	*Ctenodiscus crispatus* (Retzius)	New Brunswick Canada, NW Atlantic	NHM Zoology 1961.4.12.6-15
Porcellanasteridae	*Styracaster chuni* Ludwig	Rodrigues, Indian Ocean	NHM EE 13562
Benthopectinidae	*Benthopecten simplex* (Perrier)	Rockall Trough, NE Atlantic	NHM EE 13563
Goniasteridae	*Mediaster aequalis* Stimpson	Puget Sound, British Columbia Canada	NHM EE 13564
Archasteridae	Archaster lorioli Sukarno and Jangoux	Flic en Flac Mauritius	NHM EE 13565
Odontasteridae	*Odontaster validus* Koehler	Lecuyer Point, Antarctica	NHM Zoology 1950.1.6.34
Ophidiasteridae	*Nardoa variolata* (Lamarck)	Flic en Flac Mauritius	NHM EE 13566
Oreasteridae	*Protoreaster nodosus* (Linnaeus)	Kuta, Bali, Indonesia	NHM EE 13567
Poraniidae	*Porania pulvillus* O. F. Muller	E. Greenland, NE Atlantic	NHM Zoology 1966.1.13.27-31
Asteropseidae	*Asteropsis carinifera* (Lamarck)	Philippines	NHM EE 13568
Asterinidae	*Asterina gibbosa* (Pennant)	West Wales, NE Atlantic	NHM EE 13569
Echinasteridae	*Echinaster purpureus* (Gray)	Flic en Flac Mauritius	NHM EE 13570
Acanthasteridae	*Acanthaster planci* (Linnaeus)	Black River, Mauritius	NHM EE 13571
Solasteridae	*Crossaster papposus* (Linnaeus)	Essex coast, North Sea	NHM Zoology Unreg.
Korethrasteridae	*Remaster gourdoni* Koehler	Falklands, SW Atlantic	NHM Zoology 1948.3.16.546
Pterasteridae	*Pteraster pulvillus* Sladen	Greenland Sea	NHM Zoology 1969.6.12.87-96
Zoroasteridae	*Zoroaster fulgens* Thomson	Rockall Trough NE Atlantic	NHM EE 13572
Heliasteridae	*Heliaster helianthus* (Lamarck)	Galapagos	NHM Zoology (18)99.5.2.25-29
Asteriidae	*Asterias rubens* Linnaeus	Essex coast, UK North Sea	NHM Zoology Unreg.
Freyellidae	*Freyella elegans* (Verrill)	Rodrigues, Indian Ocean	NHM EE 13573
Brisingidae	*Brisinga costata* Verrill	Los Roques, Venezuela, Atlantic	USNM E20948
Korethrasteridae	*Korethraster hispidus* Thomson	Faroes	Nat Hist. Museum Torshavn, Faroes, 739
	Peribolaster biserialis Fisher	Bering Sea	USNM 31732
Pterasteridae	*Pteraster myonotus* Fisher	Philippines	NHM EE 13574
	Diplopteraster verrucosus (Sladen)	Tierra del Fuego, Argentina	USNM 1082940
Benthopectinidae	*Cheiraster gazellae* (Studer)	Pacific	USNM E9702
	Cheiraster sp.	Rodrigues, Indian Ocean	NHM EE 13575
	Pontaster tenuispinus (Duben and Koren)	Rockall Trough ne Atlantic	NHM EE 13576
Zoroasteridae	*Myxoderma sacculatum ectenes* Fisher	ne Pacific	USNM 32451
Astropectinidae	*Proserpinaster neozelandicus* (Mortensen)	New Zealand	NIWA Z 10882
Goniasteridae	*Pseudarchaster parelii* Duben and Koren	Alaska	USNM 32169

TABLE 1. Continued.

Family	Genus, species	Locality	Collection
Ceramaster granularis (Retzius)	North Sea	NHM EE 13577	
Caymanostellidae	*Caymanostella spinimarginata* Belyaev	East of Cuba	USNM E 20983
Pterasteridae	*Diplopteraster verrucosus* (Sladen)	Tierra del Fuego, Argentina	USNM 1082940

The first 25 taxa were used in the cladistic analysis.

fully-grown form. Radial (or carinal) abactinal ossicles extend along the radius from the primary circlet to the terminal ossicle at the tip of the arm. On the actinal surface, a central peristome is surrounded by the mouth frame (paired orals and circumorals, interradial odontophore), and ambulacral grooves run radially to the arm tips. These are constructed of opposing, paired ambulacral ossicles which alternate and articulate with spine-bearing adambulacrals. A terminal ossicle is present at the tip of each arm. The interradial areas (triangular region between the marginals, adambulacrals and orals) are occupied by small actinal ossicles. Internally, a calcified interradial septum made up of numerous small, plate-like ossicles, may extend from the mouth frame to the ambital and abactinal walls.

General morphology

Most asteroid species have five arms, but multiarmed species are represented by *Brisinga costata* Verrill, 1884 (12), *Freyella elegans* (Verrill, 1884) (11), *Heliaster helianthus* (Lamarck, 1816) (20), *Acanthaster planci* (Linnaeus, 1758) (21) and *Crossaster papposus* (Linnaeus, 1758) (12). The development of interradial arcs is very variable in the neoasteroids; these are broad and gently rounded in *Mediaster aequalis* Stimpson, 1857, *Protoreaster nodosus* (Linnaeus, 1758), *Asteropsis carinifera* (Lamarck, 1816), *Odontaster validus* Koehler, 1906, *Porania pulvillus* (O. F. Müller, 1776), *Asterina gibbosa* (Pennant, 1777), *Remaster gourdoni* Koehler, 1912 and *Pteraster pulvillus* (M. Sars, 1861). The interareas are narrow, small and acute in most other species, and absent entirely in *Freyella elegans* and *Brisinga costata*. The cross-section of the arm (Text-fig. 1) is subcylindrical in *Calliasterella mira* (Trautschold, 1879), *Asterias rubens* Linnaeus, 1758, *Zoroaster fulgens* Wyville Thomson, 1873, *Radiaster tizardi* (Sladen, 1882), *Nardoa variolata* (Risso, 1826), *Acanthaster planci*, *Echinaster purpureus* (Gray, 1840), *O. validus* and *C. papposus*; flattened and oval/rectangular in *Luidia ciliaris* (Philippi, 1837), *Astropecten irregularis* (Pennant, 1777), *Ctenodiscus crispatus* (Retzius, 1805), *Styracaster chuni* Ludwig, 1907, *Benthopecten simplex* (Perrier, 1881), *Mediaster aequalis*

and *Archaster lorioli* Sukarno and Jangoux, 1977, and abactinally convex, actinally flattened, in *Asteropsis carinifera*, *Remaster gourdoni* and *Pteraster pulvillus*. The abactinal surface is flat in *Luidia ciliaris*, *Remaster tizardi*, *Astropecten irregularis*, *Ctenodiscus crispatus*, *Styracaster chuni*, *Benthopecten simplex*, *Mediaster aequalis*, *Archaster lorioli*, *Odontaster validus* and *Porania pulvillus*, but domed in *Protoreaster nodosus*, *Asteropsis carinifera*, *Asterina gibbosa*, *Remaster gourdoni* and *Pteraster pulvillus*.

Spines

Various authors have used asteroid spines in taxonomy. For example, Fisher (1911, 1919*a*, *b*) and Koehler (1920) provided numerous drawings and photographic figures (taken with a light microscope) of dissociated spines of diverse taxa and used these in the characterization of species. The advent of the SEM has revolutionized the investigation of echinoderm stereom (e.g. Roux 1970, 1971; Smith 1980), and its application to asteroid spines provides new insight into their taxonomic and phylogenetic significance (Pl. 2).

The construction of asteroid spines shows significant and consistent variation amongst the taxa investigated. In one group, including *Luidia ciliaris*, *Astropecten irregularis*, *Ctenodiscus crispatus*, *Benthopecten simplex*, *Nardoa variolata*, *Mediaster aequalis*, *Protoreaster nodosus*, *Acanthaster planci* and *Asteropsis carinifera* (Pl. 2, figs 7–9, 13–15), spines are constructed of fine, even-sized labyrinthic trabeculae that pass distally into short to moderately elongated, fasciculate trabeculae of similar width. In some taxa (e.g. *Luidia ciliaris*, Pl. 2, fig. 7; *Benthopecten simplex*, Pl. 2, fig. 13), these terminate in short thorns, which are absent in *Nardoa variolata*, *Mediaster aequalis*, *Protoreaster nodosus* and *Acanthaster planci*. In *Echinaster purpureus*, *Asterias rubens* and *Heliaster helianthus*, the terminal thorns are enlarged, divergent and prominent (Pl. 2, figs 9, 14, 18) and are much larger and conspicuous on the abactinal spines. A strikingly different construction is present in *Crossaster papposus*, *Asterina gibbosa*, *Pteraster pulvillus*, *Remaster gourdoni*, *Porania pulvillus*, *Brisinga costata* and *Freyella elegans*, in which the shaft of both

TABLE 2. Abbreviations for asteroid morphology used in this paper.

ab	Abactinal ossicle
abr	Abactinal ridge (on amb)
abtam	Abactinal transverse amb muscle
abiim	Abactinal interradial interoral muscle (oral)
aciim	Actinal interradial interoral muscle
act	Actinal ossicle
actam	Actinal transverse amb muscle
actf	Actinal face (of oral)
actin	Actinostome
actr	Actinal ridge (on amb)
ad	Adambulacral ossicle
ada1	Single distal amb–adamb articulation
ada1a	Distal adradial amb–adamb articulation
ada1b	Distal abradial amb–adamb articulation
ada2	Proximal adradial amb–adamb articulation
ada3	Proximal abradial adamb–amb or adamb–adamb articulation
adada	Adamb–adamb articulation
adadm	Interadambulacral muscle
adca	Adoral carina
adex	Adambulacral extension
adexa	Adambulacral extension articulation
adexm	Adambulacral extension muscle
adpm	Adamb prominence (on adamb)
adr	Adradial ossicles
ads	Adambulacral spine
adsm	Adambulacral spine membrane (pterasterids)
al	Alveolus (pedicellariae)
alm	Actinolateral membrane (pterasterids)
als	Actinolateral spine base (pterasterids, korethrasterids)
amb	Ambulacral ossicle
ambb	Base of ambulacral ossicle
ambg	Ambulacral groove
ambh	Head of ambulacral ossicle
ambsh	Shaft of ambulacral ossicle
amn	Ampullar notch (on amb)
amp	Ampulla
ap	Aperture in pterasterid actinolateral membrane
apo	Apophyse on oral
aps	Apertural spine in pterasterids
art	Articulation surface on base of ped valves
bp	Basal piece (pedicellariae)
ca	Articulation on odontophore of chevron ossicles
can	Abactinal canopy in Pterasteridae
ce	Centrale
cha	Chevron ossicles arctic with odontophore of Pterasteridae
coh	Circumoral head
co	Circumoral ossicle
crib	Cribriform organ
cup	Cupule (pedicellaria)
dab	Distal abductor (crossed peds)
dad	Distal adductor (crossed peds)
dadam	Distal amb–adamb muscle
dcoa	Distal circumoral articulation on oral
dcp	Distal circumoral process on circumoral
de	Dentition (orals, ambs, peds)

TABLE 2. Continued.

df	Distal flange (pterasterid and korethrasterid ambs)
doda	Distal odontophore articulation (on oral, odontophore)
exab	External abductor (pedicellariae)
exod	External face of odontophore
fs	Attachment of furrow spine
ico	Interradial chevron ossicles
ig	Interradial groove
ilad	Internal longitudinal adductor (pedicellariae)
im	Inferomarginal
ima	Inferomarginal articulation on axillary/ odontophore
int	Intermarginal
iioa	Interradial interoral articulation (on oral)
is	Interradial septum
itad	Internal transverse adductor (pedicellariae)
k	Keel (on odontophore)
ka	Keel articulation on axillary/oral in *C. mira*
lia	Longitudinal interamb articulation
lim	Longitudinal interambulacral muscle
ln	Lateral notch (on odontophore)
ls	Lateral surface of keel (on odontophore)
mad	Madreporite
od	Odontophore
od/ax	Odontophore/axillary
odc	Odontophore muscle capsule (on oral)
odom	Oral-odontophore muscle
or	Oral ossicle
orada	Adambulacral articulation (on oral)
oradm	Oral adambulacral muscle
osp	Attachment of oral spine
pab	Proximal abductor (pedicellaria)
padam	Proximal adamb–amb muscle
pb	Proximal blade (oral ossicle)
pcoa	Proximal oral-circumoral articulation
pcp	Proximal circumoral process (on circumoral)
pe	Peristome
ped	Pedicellaria
pedu	Peduncle (on pedicellaria)
pir	Primary interradial ossicle
pn	Podial notch (on adamb)
poda	Proximal odontophore articulation (on oral and odontophore)
pr	Primary radial
ra	Radial
rart	Radial articulation of proximal blade (on oral)
riom	Radial interoral muscle
rng	Ring nerve groove on oral
rvg	Ring vessel groove on oral
sa	Superambulacral articulation on amb base
samb	Superambulacral
sm	Superomarginal ossicle
sos	Attachment of suboral spine
t	Terminal ossicle
tb	Transverse bar of odontophore
tf	Tube foot
1st tf	Pace occupied by first tube foot between oral/ circumoral
va	Valves of pedicellaria

TABLE 3. Comparison of morphological terms for adambulacrals.

Turner and Dearborn (1972)	Blake (1973)	Gale herein
Aboral alveolus	UG	actam
Dentition	dn	de
Articular surface of articulation complex	dN,pN	lia
Alveolus of articular complex	2G	lim
Capitulum	Amb body	ambh
Shaft	Amb body	ambsh
Base	Amb extension	ambb

TABLE 4. Comparison of morphological terms for ambulacrals.

Turner and Dearborn (1972)	Blake (1973)	Gale herein
Prox alveolus of aboral face	Pm1	padam
Distal alveolus of aboral face	Pm2	dadam
Alveolus of distal side face	Pm3	adadm
Alveolus of proximal side face	Pm3	adadm
Pitted articulation of aboral face	Pa1	ada1
Abradial articulation of aboral face	Pa4	ada3
Depressed articulation on surface of proximal side face	Pa3	adada
Adradial projection of adoral face	Pa2	ada2

adambulacral and abactinal spines is constructed of smooth, glassy fasciculate trabeculae parallel to the long axes of the spines; these are conjoined by short transverse struts (Pl. 2, figs 1–3, 5–6, 10–12, 16). The terminations of the abactinal spines form lance-like spikes in *Crossaster papposus*, *Remaster gourdoni* and *Asterina gibbosa* (Pl. 2, figs 10, 12, 16). The abactinal spines are low, closely artic-

ulating polygonal granules in *Mediaster aequalis*, *Protoreaster nodosus* and *Nardoa variolata* (Pl. 2, fig. 4).

Calcified interradial septa

Calcified interradial septa found in neoasteroids comprise two discrete structural types. In one, seen in *Asterina gibbosa* (Pl. 1) and *Crossaster papposus*, the septum comprises a pillar of small ossicles that extends from the distal abactinal surface of the odontophore to the inner surface of the abactinal disc. In the other, the septum forms a more or less entire wall of ossicles which separate the coelomic spaces of each radius. This wall attaches to the odontophore, the actinal and abactinal inner surface of the disc, and appears to be partially formed of occluded actinal and abactinal ossicles. This is found in *Protoreaster nodosus*, *Acanthaster planci*, *Asterias rubens*, *Zoroaster fulgens* and *Heliaster helianthus*.

Chevron ossicles and interradial grooves

The presence of distinctive interradial grooves floored by paired columns of ossicles in korethrasterids and pterasterids (*Pteraster pulvillus*, *Remaster gourdoni*; Text-fig. 2A–C) has been known since early descriptions (e.g. Danielssen and Koren 1884). These comprise V-shaped, paired, bar-like ossicles that articulate with the odontophore proximally, here called chevron plates, and distally with ossicles interpreted by Fisher (1940) as inferomarginals in *Peribolaster folliculatus* Sladen, 1889. The chevron ossicles floor narrow interradial grooves, which extend abactinally to the primary radial ossicles (*Remaster gourdoni*; Text-fig. 2). These plates are homologous with the ossicles of the calcified interradial septa.

TABLE 5. Comparison of morphological terms for mouth frame ossicles.

Turner and Dearborn (1972)	Blake (1973)	Gale herein
Pustules of the sutural spines	Sp b	sos
Dentition	dn	de
Alveolus of distal side face		oradm
Articular surface for 2nd adambulacral		orada
Articular surface for distal jaw process of odontophore		doda
Articular surface for proximal jaw process of odontophore		Proximal odontophore articulation – poda
Alveoli for attachment of fibres from odontophore		Odontophore-oral muscle – odom
Fibres connecting members of jaw		Abactinal interradial interoral muscle – abiim
Fibres for the proximal attachment of the opposing members of a jaw		Actinal interradial interoral muscle – aciim
Adradial surface of apophyse connecting opposite members of adjacent jaws		Radial interoral muscle – riom

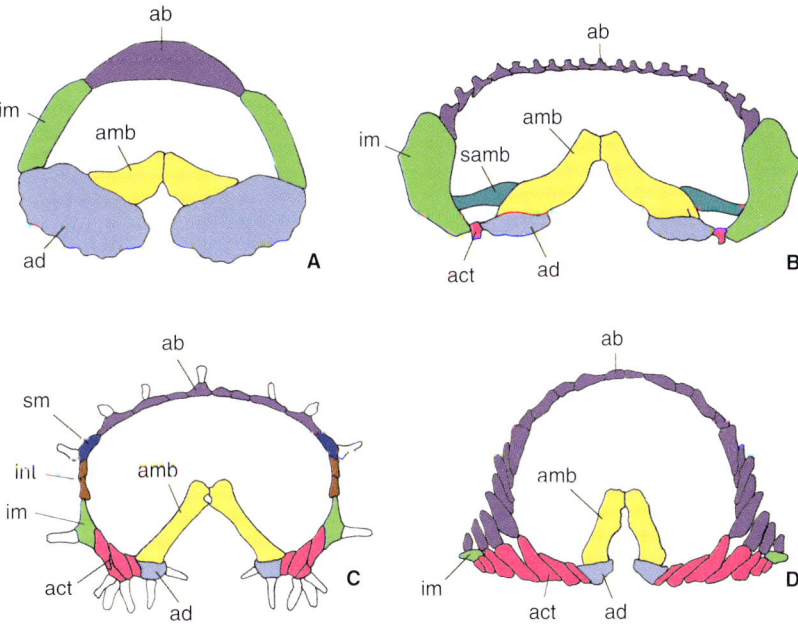

TEXT-FIG. 1. Cross-section of the arms of: A, *Calliasterella mira*. B, *Luidia ciliaris*. C, *Asterina gibbosa*. D, *Asterias rubens*. Note the very broad adambulacrals in *C. mira*. The cross-section of the arm is classified as subcylindrical in A and C, subrectangular in B, and convex-flat in C. For abbreviations, see Table 2.

Abactinal ossicles

Abactinal ossicles are highly diverse in asteroids, and reference is made to Blake (1987) and Clark and Downey (1992) for illustrations of different types. The retention of enlarged primary abactinals ossicles (primary radials, interradials, centrale, single radial row), a characteristic of juvenile asteroids, into adult species occurs widely amongst both Palaeozoic and post-Palaeozoic taxa (Kesling and Strimple 1966; Spencer and Wright 1966; Downey 1970). Other taxa have smaller, more uniformly sized and shaped abactinals. Some are paxilliform, with a broader base, often flanged to articulate with adjacent ossicles and to allow the passage of papulae, and a tall pedicel, the abactinal termination of which bears spines. Although there have been attempts to distinguish between different types of paxillae (megapaxillae, etc.), all have a basically similar structure. Tabular abactinals are low and broad, and the spines carried are short and granular; with increasing differentiation of a pedicel, these grade into paxillae. Abactinals may be low and conical or possess several flanges, which imbricate with adjacent ossicles and articulate by means of tongue-and groove articulations. Abactinals can be tessellate or imbricate. They can carry numerous spines, or a single central spine, which may have a central concave articular structure.

Madreporite

A single abactinally positioned madreporite is present in all taxa, except for *A. planci* in which numerous madreporites are present. The position of the madreporite on the disc displays considerable variation. The ratio of the distance between the inner margin of the superomarginal and the distance between the centrale and madreporite (PM) gives values of >1 for *Luidia ciliaris*, *Styracaster chuni*, *Ctenodiscus crispatus*, *Astropecten irregularis*, *Calliasterella mira*, *Radiaster tizardi* and *Benthopecten simplex*, and <1 for all other taxa, except in *Brisinga costata* and *Freyella elegans* in which PM cannot be measured.

Marginals

The size and prominence of marginal ossicles has long been used in asteroid classification, and Sladen (1889) based his fundamental dichotomy of the asteroids upon the relative development of these ossicles (Cryptozonida, Phanerozonida). Marginal ossicles vary very widely in morphology, from large block-like structures, which form a conspicuous border to the ambitus (*Luidia ciliaris*, Text-fig. 1B; *Astropecten irregularis*, *Archaster lorioli*, *Mediaster aequalis*), tiny inconspicuous ossicles (*Asterina gibbosa*; Text-fig. 1C; Pl. 1) to paxilliform ossicles (*Crossaster papposus*, *Odontaster validus*). Only a single marginal row is present in *Calliasterella mira*, *Zoroaster fulgens* (see Blake 1987; Blake and Elliot 2003) and *Luidia ciliaris*; in all other neoasteroids, discrete infero- and superomarginals are present. Actinal–abactinal grooves between marginals, carrying small ciliated spines, are called intermarginal fascioles and are found in *Luidia ciliaris*, *Astropecten irregularis*, *Styracaster chuni* and *Ctenodiscus crispatus*. In the two last-named species, these carry spine-supported vertical epithelial folds called cribriform organs. In *Remaster gourdoni* and *Pteraster pulvillus*, marginals are not present.

Actinals

Actinal ossicles are not present in *Calliasterella mira* and most other Palaeozoic asteroids (Blake and Hotchkiss 2004; Shackleton 2005), and the axillary ossicle, a variably occluded single interradial ossicle, usually is in contact with the oral ossicles, and actinal interareas are absent (Gale 1987; Blake and Hotchkiss 2004). The odontophore is homologous with the axillary and is entirely or largely internal in neoasteroids (Gale 1987; Pl. 1 herein). Most neoasteroids possess triangular actinal interradial areas (interareas), which extend between the orals, adambulacrals and inferomarginals made up of small flat polygonal or subrounded ossicles (Pl. 1, fig. 4). The interareas are not discernible in *Brisinga costata*, *Freyella elegans*, *Remaster gourdoni* and *Pteraster pulvillus*. The actinal ossicles are arranged in two separate patterns in neoasteroids (Hotchkiss and Clark 1976), which Blake and Hotchkiss (2004) characterized as the Ambulacral Column Pattern (ACP) and the Marginal Row Pattern (MRP), respectively. In the ACP, the longest column of actinals is parallel with the ambulacral column; the smallest and most irregularly arranged actinals are adjacent to the marginals, where new plates are added during growth (Pl. 1, fig. 4). By contrast, in the MCP condition, the longest row is adjacent to the marginal ossicles, and new plates are added along the margin of the adambulacrals. The majority of taxa investigated here fall in the ACP category. The porcellanasterids were described as possessing an MRP pattern by Blake and Hotchkiss (2004), but in *Styracaster chuni* new plates are added adjacent to the marginals (e.g. Madsen 1961, fig. 20), although the longest and largest actinal row is adjacent to the adambulacrals. The MRP condition is present in *Porania pulvillus*, *Zoroaster fulgens* and *Asterias rubens*. An irregular mosaic pattern with small plates introduced interstitially is seen in *Protoreaster nodosus*, which is presumably a modified form of the ARP arrangement.

Actinal ossicles are arranged in well-defined single or double rows that extend from the marginals to the adambulacrals in *Radiaster tizardi*, *Ctenodiscus crispatus*, *Odontaster validus*, *Porania pulvillus* and *Asterina gibbosa* (Pl. 1, fig. 4). Fasciolar channels running from the groove to the intermarginal fascioles are inset into these rows in *Ctenodiscus crispatus* and *Radiaster tizardi*.

Pedicellariae

Although pedicellariae have been widely used in asteroid classification since Perrier's work in the late nineteenth century, there have been rather few descriptions of their morphology and comparative anatomy, with the exception of studies on forcipulate pedicellariae found in the forcipulatid family Asteriidae (Chia and Amerongen 1975; Lambert *et al.* 1984; Roberts and Campbell 1988). A review of all asteroid pedicellariae by Jangoux and Lambert (1987) provided a valuable classification of such appendages, but did not address the question of homology between types of pedicellariae and their component structures. Indeed, many authors who discussed pedicellariae in asteroids have presumed that considerable convergence and independent evolution has taken place (e.g. Jangoux and Lambert 1987; Blake and Reid 1998). The present study has concluded that there is compelling evidence for the existence of homologies between elementary, alveolar and complex pedicellariae.

Elementary pedicellariae (Type A of Jangoux and Lambert 1987) comprise two to four (rarely more) similar valves, which have a nonspecialized articulation upon a primary ossicle. They are commonly present and highly variable in the genus *Luidia* Forbes, 1839 (see Clark and Downey 1992, fig. 7), and two-, three- or four-valved forms can be present in the same species. The small three-valved forms (which resemble tiny beech nuts), figured by Clark and Downey (1992), in *Luidia* are very similar in form to the smaller elementary pedicellariae seen in the Devonian multiarmed genus *Arkonaster* (Kesling 1982).

Of the asteroids covered in the present study, only *Luidia ciliaris* possesses elementary pedicellariae (Pl. 3, figs 1–4; Text-fig. 3A). These are all bivalved and similar in

---

EXPLANATION OF PLATE 1

Overall construction of the asteroid skeleton illustrated by denuded and partly dissected *Asterina gibbosa*, with central region of abactinal ossicles removed. Abactinal (1–3) and actinal (4) aspects.

Fig. 1. Enlargement of ambulacral groove.

Fig. 2. Mouth frame, enlarged.

Fig. 3. Abactinal view of partly dissected skeleton, to show mouth frame, interradial ossicles and ambulacral groove ossicles.

Fig. 4. Actinal aspect of same individual, to show ambulacral grooves, peristome, and actinal ossicles.

See Table 2 for abbreviations.

Scale bars represent 5 mm.

PLATE 1

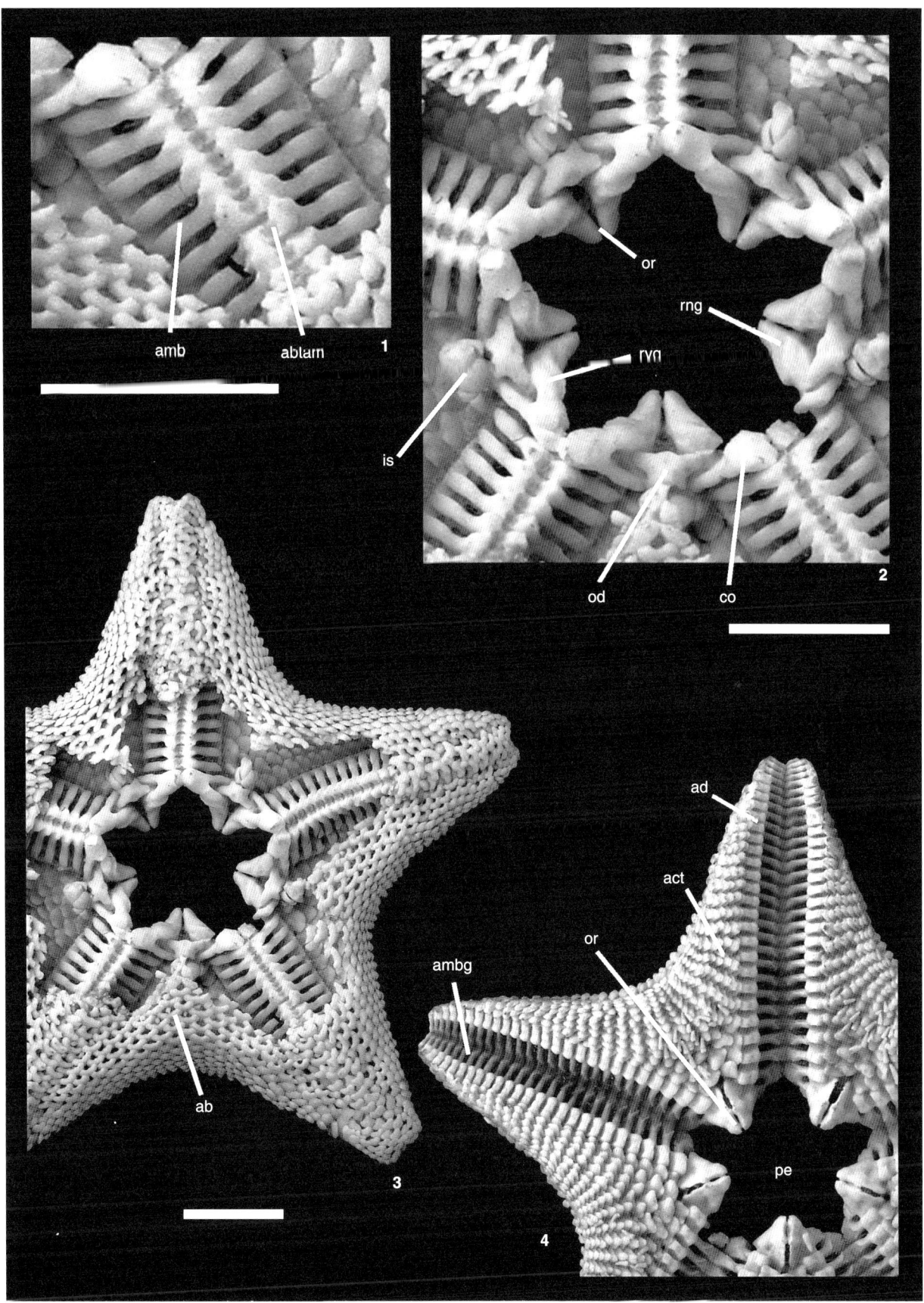

GALE, Skeletal morphology of Recent *Asterina gibbosa*

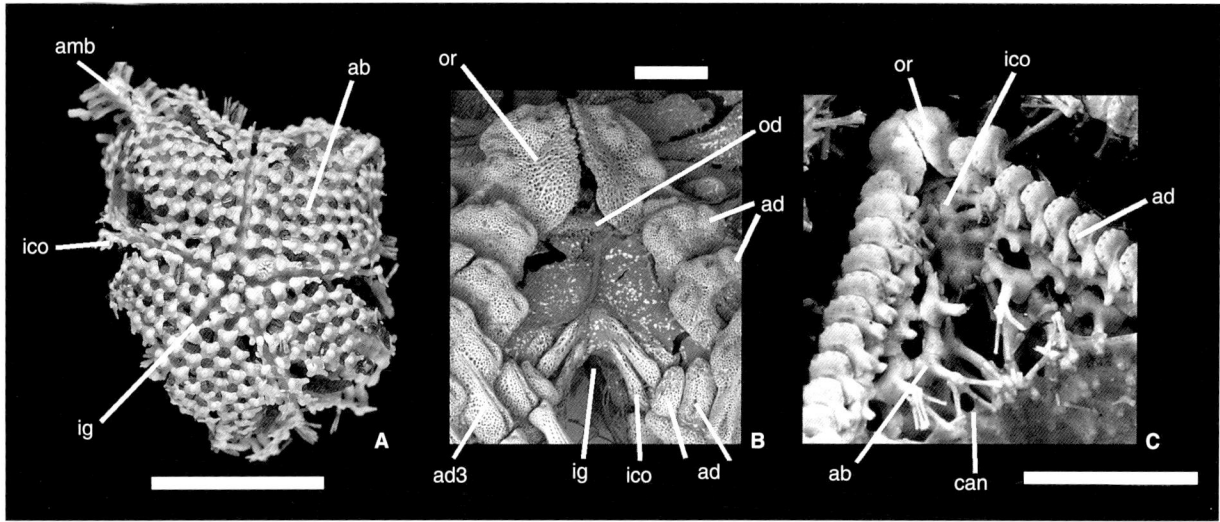

TEXT-FIG. 2. Interradial chevron ossicles and grooves in *Remaster gourdoni* (A, B) and *Pteraster pulvillus* (C). A, abactinal aspect of denuded specimen, to show shallow, narrow, but very well-defined interradial grooves. B, actinal interradial view of same specimen, showing interradial chevron plates. C, interradial view of denuded *Pt. pulvillus*, to show imbricated interradial chevron plates, abactinal megapaxillae and canopy. See Table 2 for abbreviations. Scale bars in A, C represent 10 mm; B, 1 mm.

form, and a single one is present on each adambulacral and each actinal intermediate ossicle. The two valves articulate with each other at the base and are adducted by an internal transverse muscle (*itad*; Pl. 3, figs 3–4; Text-fig. 3A). The pedicellariae attach on the surface of an ossicle and are abducted by two small external muscles (*exab*) that run from the base of the valves to the underlying ossicle (Pl. 3, figs 1, 4; Text-fig. 3A). The surface of the ossicle to which the pedicellaria is attached is not specially modified, and neither an alveolus nor a specialized articulation surface is present (Pl. 3, fig. 2). In other species of *Luidia*, more than one pedicellaria can be present on a single ossicle, usually an adambulacral (Clark and Downey 1992). Elementary pedicellariae were recorded in *Archaster* Müller and Troschel, 1840 and *Acanthaster planci* (Lamarck (1758) by Jangoux and Lambert (1987), but were not found in these taxa by the present author in spite of an extensive search.

Bivalved alveolar pedicellariae (Type B of Jangoux and Lambert 1987) are found in *Archaster lorioli* (Pl. 3, figs 5–9; Text-fig. 3B) and are essentially similar in construction to elementary pedicellariae in that the two valves articulate together at the base and two external abductors (*exab*) and one internal transverse adductor (*itad*) are

SEM illustrations of spines in diverse asteroid taxa.

Fig. 1. *Pteraster pulvillus*, actinolateral spine from lateral extremity of adambulacral ossicle.

Figs 2, 16. *Asterina gibbosa*, adambulacral and abactinal spines, respectively.

Figs 3, 12. *Crossaster papposus*, adambulacral and abactinal spines, respectively.

Figs 4, 17. *Nardoa variolata*, abactinal and adambulacral spines, respectively.

Fig. 5. *Porania pulvillus*, inferomarginal spine.

Fig. 6. *Brisinga costata*, adambulacral spine.

Fig. 7. *Luidia ciliaris*, adambulacral spine.

Figs 8, 18. *Echinaster purpureus*, adambulacral and abactinal spines.

Fig. 9. *Heliaster helianthus*, abactinal spine.

Fig. 10. *Remaster gourdoni*, lateral adambulacral spine.

Fig. 11. *Odontaster validus*, abactinal spine.

Fig. 13. *Benthopecten simplex*, superomarginal spine.

Fig. 14. *Asterias rubens*, abactinal spine.

Fig. 15. *Acanthaster planci*, adambulacral spine.

Scale bar represents 1 mm.

PLATE 2

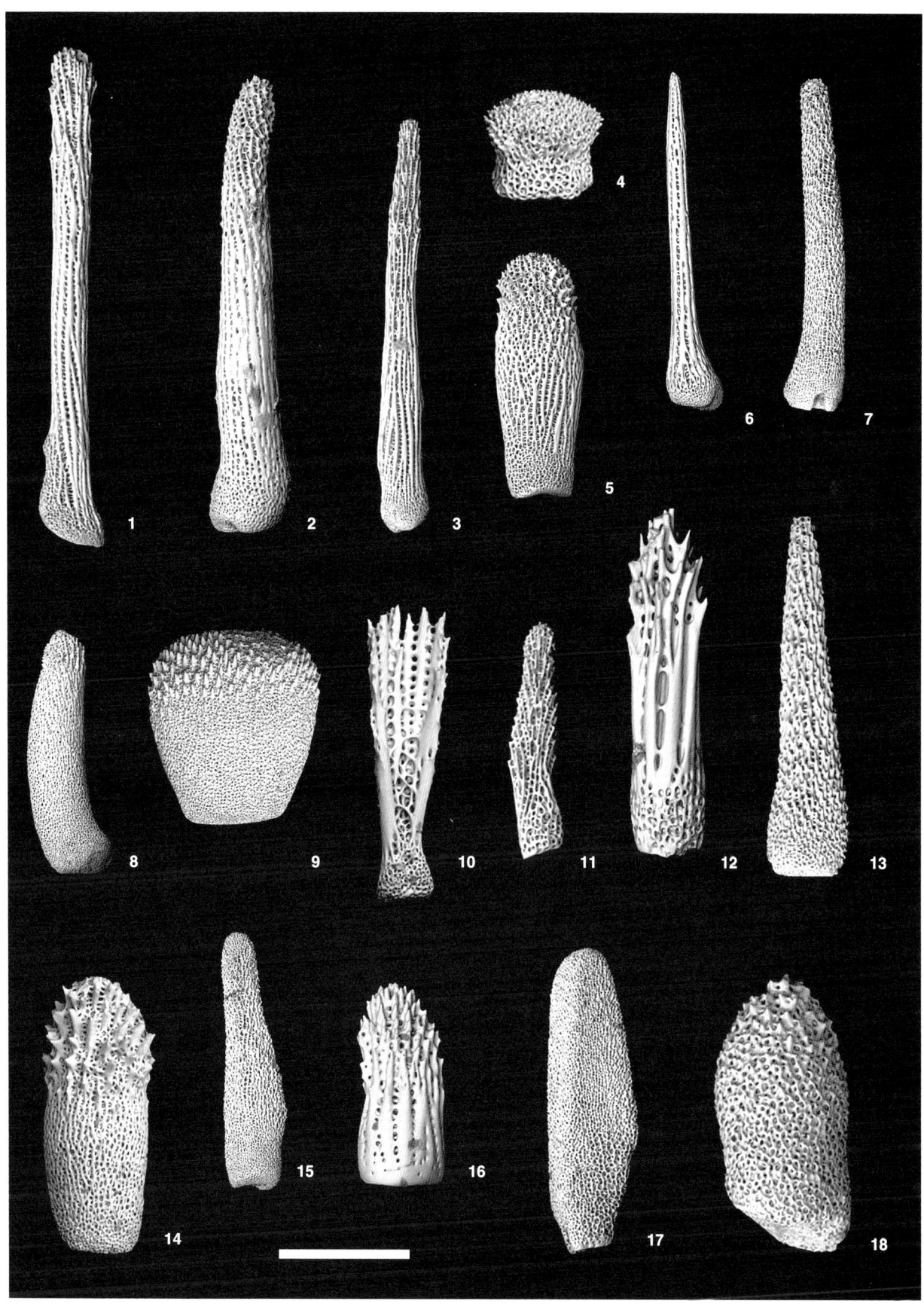

GALE, Morphology of Recent asteroid spines

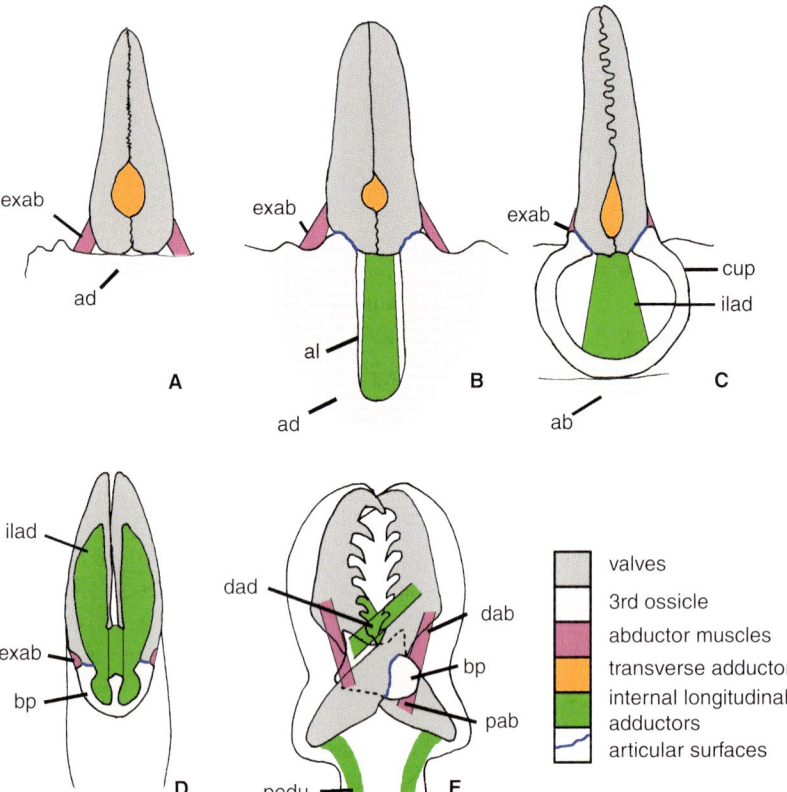

TEXT-FIG. 3. Inferred homologies of neoasteroid pedicellariae. A, Elementary, *Luidia ciliaris*. B, C, Alveolar, *Archaster lorioli*; *Acanthaster planci*. D, E, Forcipulate; D, *Marthasterias glacialis*, straight ped, modified after Lambert *et al.* (1984, fig. 6); E, *Stylasterias forreri*, crossed ped, after Chia and Amerongen (1975, fig. 11). See Table 2 for abbreviations.

present (Pl. 3, figs 5–6, 8; Text-fig. 3B). However, each pedicellaria rests upon a deep, parallel-sided blind cavity (an alveolus; Pl. 3, figs 7, 9; Text-fig. 3B) set in the underlying (primary, usually adambulacral) ossicle. The external part of the alveolus (Pl. 3, figs 7, 9) is specialized for attachment of the base of the pedicellaria, with two small regions of smooth perforate stereom for articulation of the valves and two raised 'lips' that extend the articulation surfaces and provide sites for insertion of the two external abductors. Two internal adductor muscles (*ilad*) originate on the base of the alveolus and one inserts onto the base of each valve. Only one pedicellaria is ever found upon each basal ossicle. The incorporation of a basal ossicle as an integral component of each pedicellaria is here

identified as a fundamental and unique evolutionary innovation and the basal ossicle with an alveolus is homologous with the basal piece of complex (forcipulate) pedicellariae (Text-fig. 3; Jangoux and Lambert 1987).

In *Mediaster aequalis* and *Archaster lorioli*, bivalved alveolar pedicellariae are found only upon primary ossicles, but in *Acanthaster planci* and *Protoreaster nodosus* these are also found attached to small, delicate, cupshaped secondary ossicles (here termed cupules), which are embedded superficially in the dermal structure (Pl. 4, figs 1–3; Text-fig. 3C). Cupules possess specialized articulation surfaces for the valves (Pl. 4, figs 1–2) and the paired internal adductors (*ilad*) insert on their basal interior region (Pl. 4, fig. 3). Although these structures have

---

EXPLANATION OF PLATE 3

SEM images of pedicellariae and associated ossicles.

Figs 1–4. *Luidia ciliaris*. 1, 3, valve of pedicellaria in external and oblique internal views, respectively; 2, attachment site of a pedicellaria on an adambulacral; 4, proximal view of bivalved pedicellaria.

Figs 5–9. *Archaster lorioli*. 5–6, oblique external view of pedicellaria valve base and internal aspect of same, respectively; 7, external view of attachment site of pedicellaria on adambulacral; 8, lateral aspect of pedicellaria; 9, broken surface of adambulacral to show alveolus.

See Table 2 for abbreviations.

Figs 1–4, ×100; Figs 5, 6, ×125; Figs 7, 8, ×70; Fig. 9, ×35.

PLATE 3

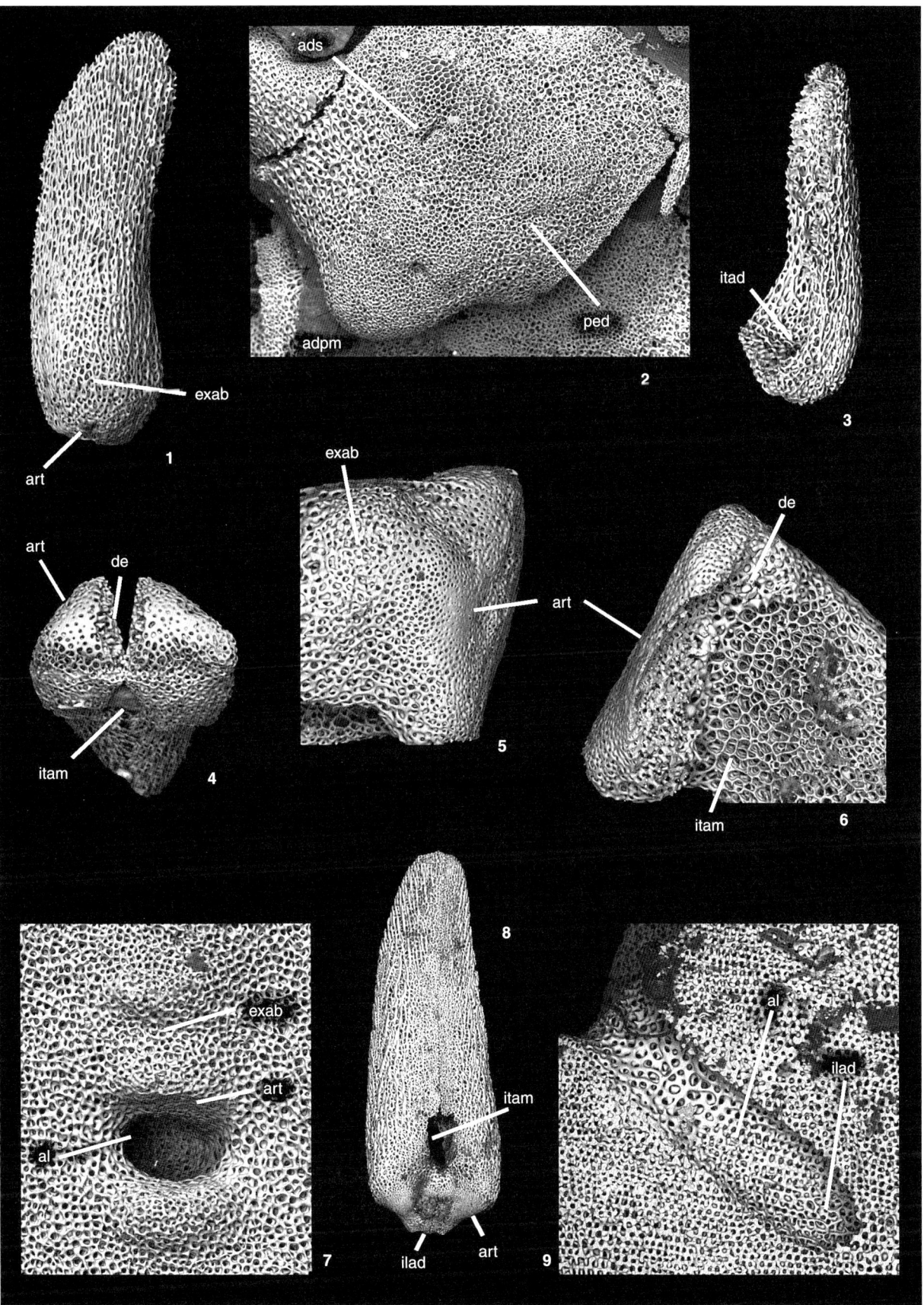

GALE, Pedicellariae of Recent asteroids

been occasionally figured and commented upon (the 'schüsselförmigen Basalstücke', bowl-shaped basal pieces, of Döderlein 1916), they have not hitherto been discussed in the context of their possible homologies. Essentially, cupules represent an intermediate morphological stage between the large basal ossicle of alveolar pedicellariae and the small basal piece of straight forciculate pedicellariae (Text-fig. 3C, D). It is interesting that many alveolar pedicellariae contain an internal process, referred to as 'cone ligamentaire' (ligament cone) by Jangoux and Lambert (1987, fig. 8), which is homologous with an internal process on the basal piece of straight forciculate pedicellariae to which the inner adductor muscles are attached (Text-fig. 3D).

Forciculate pedicellariae (complex pedicellariae or Type C of Jangoux and Lambert 1987) are the most distinctive and best-studied type of asteroid pedicellariae and are found in *Asterias rubens*, *Zoroaster fulgens*, *Heliaster helianthus*, *Freyella elegans* and *Brisinga costata* amongst species studied here. They all possess a flexible, muscular peduncle by which they are attached to the surface of the asteroid. In straight forciculate pedicellariae (Pl. 4, fig. 6; Text-fig. 3D), the basal piece is cup-shaped, and the valves articulate with the upper margins of the ossicle. Two small external abductors are present as in elementary and alveolar pedicellariae. Internally, the valves are hollow, and two longitudinal adductors (inner, outer) run from each valve to the basal piece. In crossed forciculate pedicellariae, the strongly toothed valves have become crossed to form a scissor-like structure, and the basal piece is incorporated as a highly modified bridging structure where the valves cross (Pl. 4, fig. 10; Text-fig. 3E; Lambert *et al.* 1984). One pair of adductor muscles has been lost, and both proximal and distal abductors are present (Chia and Amerongen 1975). Crossed pedicellariae are present in all the forciculatid species listed above with the exception of *Zoroaster fulgens*, which possesses only straight pedicellariae. Lambert *et al.* (1984) argued that crossed forciculate pedicellariae were derived from straight forciculate pedicellariae.

Ambulacral groove

The ambulacral groove has a basically similar construction in *Calliasterella* Schuchert, 1914, and all the neoasteroid taxa examined. The flexible groove is made up of paired, opposing ambulacrals, which articulate firmly across the mid-radial line (Pls 1, 5; Text-fig. 4F) and with their proximal and distal neighbours. The external tube feet and internal ampullae pass through notches between successive ambulacrals (Text-fig. 4F). The adambulacrals alternate with the ambulacrals and articulate with the basal abradial (actinal) part of the ambulacrals (*ambb*; Text-fig. 4A). The ambulacral groove construction of most Palaeozoic asteroids differs significantly from the taxa described here. For example, in many Palaeozoic taxa, the adambulacrals and ambulacrals do not alternate, and notches to accommodate internal ampullae are absent (Gale 1987; Blake and Guensburg 1988; Shackleton 2005).

Ambulacral ossicles

The ambulacrals of all taxa in the present study are made up of three morphological and functionally discrete regions: head, shaft and base (Text-fig. 4A–C). The ambulacral head (*ambh*) articulates with the corresponding ambulacral of the opposing column across the mid-radial line by means of an interdigitating dentition (*de*; Pl. 5; Text-fig. 4A). In all taxa, except *Calliasterella mira* (Text-figs 5A, 6D–G), paired abactinal (*abtam*) and actinal transverse ambulacral muscles (*actam*) are present, positioned above and beneath the dentition, respectively. In most neoasteroid taxa, the abactinal transverse muscle is external, and its area of insertion leaves a semicircular to triangular depression on the abactinal part of the ambulacral head (Pl. 5, figs 1–7, 9; Text-fig. 5B–D). The muscles serve to open and close the ambulacral groove (Heddle 1967). However, in the brisingid *Brisinga costata* (Pl. 8, fig. 4) and the freyellid *Freyella elegans* (Pl. 5, fig. 8), *abtam* is short and internal, inset between separate

---

EXPLANATION OF PLATE 4

SEM images of pedicellariae and associated ossicles.

Figs 1–5, 7–9. *Acanthaster planci*, dissociated cupula in lateral (1), external (2) and internal (3) aspects; 5, partially cleaned cupula and basal pedicellaria; 4, internal view of single valve of pedicellaria; 7, basal view of single valve of pedicellaria; 8–9, external and internal views of basal portion of valve of pedicellaria.

Fig. 6. *Asterias rubens*, abactinal straight pedicellaria in lateral view, soft tissue partially removed.

Fig. 10. *Heliaster helianthus*, crossed pedicellaria in lateral view, soft tissue removed.

See Table 2 for abbreviations.

Figs 1–3, ×60; Fig. 4, ×35; Fig. 7, ×70; Fig. 5, ×125; Figs 8–9, ×75; Figs 6, 10, ×250.

PLATE 4

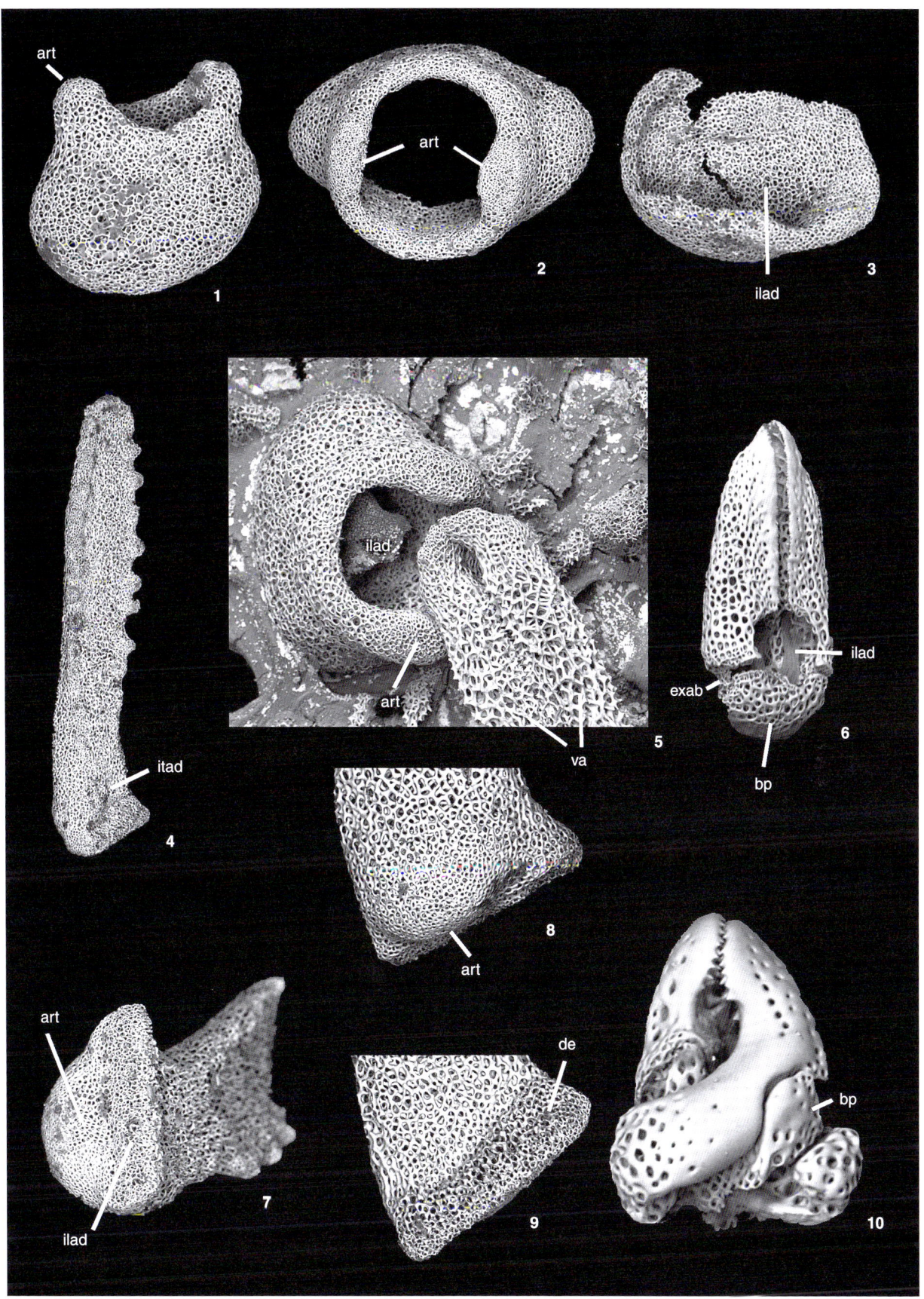

GALE, Pedicellariae of Recent asteroids

proximal and distal areas of dentition; the groove cannot be opened or closed significantly in these taxa.

The contacts between successive ambulacrals of a column are set at right angles to the mid-radial line, and the heads do not imbricate in *Calliasterella mira* (Text-fig. 5A; expressed as an ambulacral angle of about 90 degrees), *Astropecten irregularis* (Pl. 5, fig. 1; Text-fig. 5B), *Luidia ciliaris*, *Radiaster tizardi*, *Ctenodiscus crispatus* (Pl. 5, fig. 2), *Styracaster chuni* and the benthopectinids *Benthopecten simplex* and *Pontaster tenuispinus* (Düben and Koren, 1846) (Pl. 5, fig. 3). In all other neoasteroids studied, the ambulacral heads imbricate proximally to varying degrees (Pl. 5, figs 4–9; Text-fig. 5C, D), and the articulation surfaces between successive ambulacrals (*lia*) are inclined proximally. In one group of taxa (*Nardoa variolata* (Pl. 10, figs 1–2), *Protoreaster nodosus*, *Mediaster aequalis*, *Asteropsis carinifera*, *Archaster lorioli*, *Zoroaster fulgens*, *Asterias rubens* and *Heliaster helianthus*), the imbrication is only moderate, but the longitudinal interambulacral muscle (*lim*) is visible in abactinal view, and an ambulacral angle of 55–70 degrees is present (Pl. 5, figs 5–6; Text-fig. 5C). In a further group, the ambulacral head is flattened and triangular and imbricates strongly over the adjacent, proximal, ambulacral ossicle (*Crossaster papposus*, Pl. 5, fig. 7; Pl. 10, figs 3–4; Text-fig. 9B; *Asterina gibbosa*, Pl. 5, fig. 4; Text-fig. 9A; *Remaster gourdoni*, Text-fig. 9E; *Pteraster pulvillus*, Text-fig. 9D; *Porania pulvillus*, *Acanthaster planci*), giving an ambulacral angle of 30–50 degrees. In *Freyella elegans* and *Brisinga costata*, there is a slight distal inclination of the interambulacral articulation, and the ambulacral heads resemble vertebral centra (Pl. 5, fig 8). There is an additional distal wing-like flange (*df*) on the ambulacral heads of *Pteraster pulvillus* (Text-fig. 9D) and *Remaster gourdoni* (Text-fig. 9E), which carries an additional interambulacral muscle.

Ambulacral heads of all taxa possess longitudinal interambulacral articulation surfaces (*lia*) on the abradial side of which a longitudinal interambulacral muscle (*lim*) inserts (Pl. 10; Text-fig. 4). Contraction of this muscle contributes to raising the arm (Heddle 1967). The interambulacral articulation surface (*lia*) comprises a round or oval raised region of smooth perforate stereom in *Calliasterella mira*, *Astropecten irregularis*, *Luidia ciliaris*, *Styracaster chuni*, *Radiaster tizardi*, *Ctenodiscus crispatus*, *Benthopecten simplex*, *Nardoa variolata* (Pl. 7, fig. 10; Pl. 10, figs 1–2, 5–6), *Mediaster aequalis*, *Protoreaster nodosus* and *Asteropsis carinifera*. In species with a flattened, imbricating ambulacral head, the proximal articulation surface is flat and triangular and positioned on the proximal margin of the head (Pl. 10, figs 3–4; Text-fig. 9). In *Freyella elegans* and *Brisinga costata*, the interambulacral articulation structure is elongated, large and concavo-convex (Pl. 8, fig. 6). The *lim* insert onto an area immediately abactinolateral to the articulation surface. In most taxa, the proximal and distal attachment sites are similar in size and shape (e.g. *Nardoa variolata*; Pl. 7, figs 4–5; Pl. 10, figs 1–2), but in species with flattened, imbricating ambulacral heads (*Crossaster papposus*, *Asterina gibbosa*, *Remaster gourdoni*, *Pteraster pulvillus*, *Porania pulvillus* and *Acanthaster planci*), the site of insertion of the proximal *lim* is elongated and deeply inset on the proximal margin (Pl. 10, fig. 3). The enlarged, flattened to slightly concave distal muscle facet is positioned on the abactinal surface of the ambulacral head (Pl. 10, fig. 4).

The ambulacral shaft (*ambsh*) is distinctively constructed of transversely elongated glassy trabeculae in *Asterina gibbosa*, *Odontaster validus*, *Crossaster papposus*, *Remaster gourdoni* (Text-fig. 9), *Porania pulvillus* and *Pteraster pulvillus* (see also illustrations in Clark and Downey 1992). A similar construction is seen in *Asterias rubens* and *Heliaster helianthus*, species in which the ambulacrals are shortened in a proximal–distal direction (Pl. 5, fig. 9; Pl. 8, fig. 3). In other taxa, the shaft is made up of labyrinthic stereom.

EXPLANATION OF PLATE 5

Abactinal aspect of ambulacral construction in neoasteroids, to illustrate ambulacral heads and relative imbrication.

Fig. 1. *Astropecten irregularis.*

Fig. 2. *Ctenodiscus crispatus.*

Fig. 3. *Pontaster tenuispinus* (Benthopectinidae).

Fig. 4. *Asterina gibbosa.*

Fig. 5. *Ceramaster granularis* (Goniasteridae).

Fig. 6. *Nardoa variolata.*

Fig. 7. *Crossaster papposus.*

Fig. 8. *Freyella elegans.*

Fig. 9. *Asterias rubens.*

See Table 2 for abbreviations.

Scale bars represent 5 mm.

PLATE 5

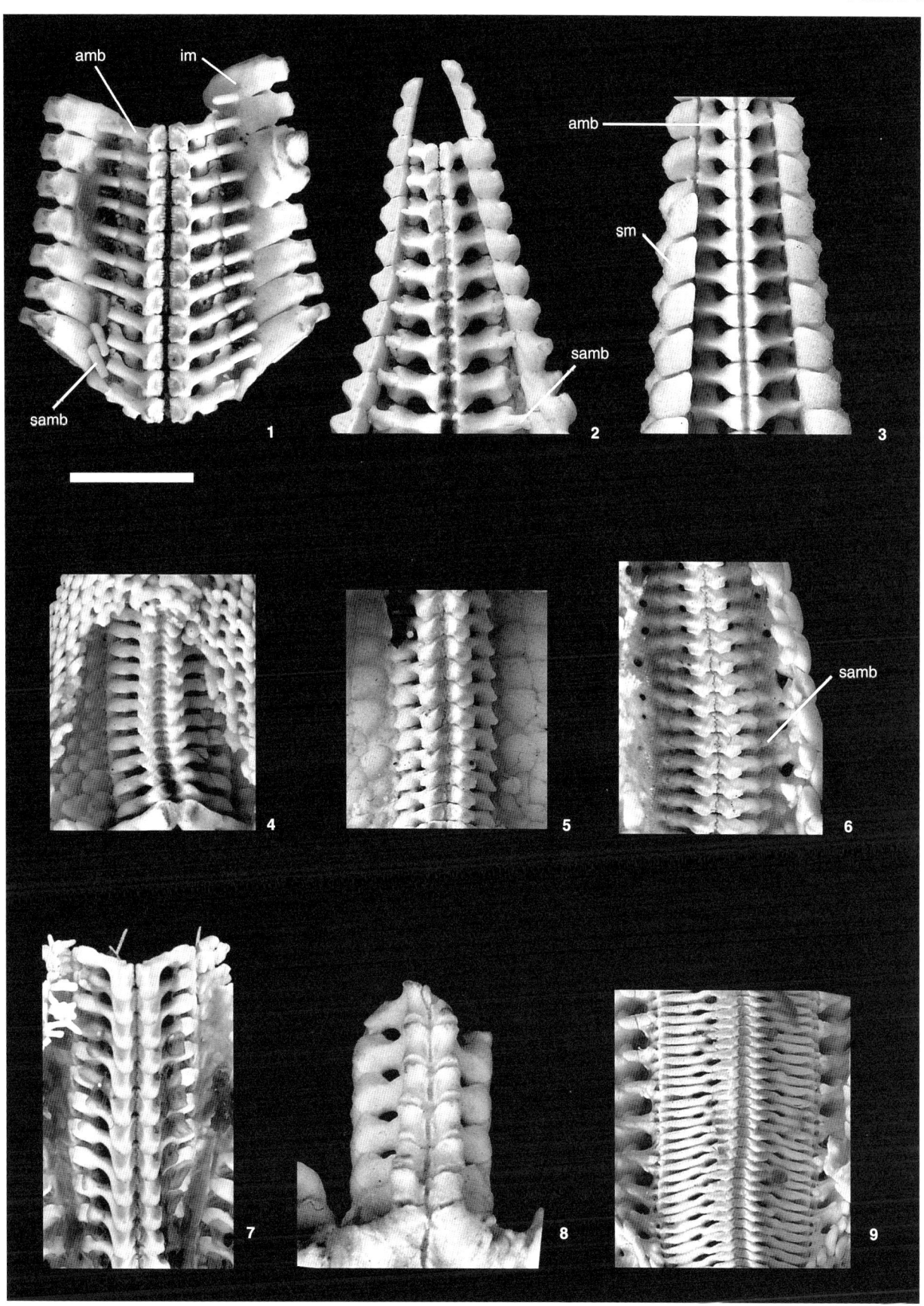

GALE, Ambulacral head morphology in Recent asteroids

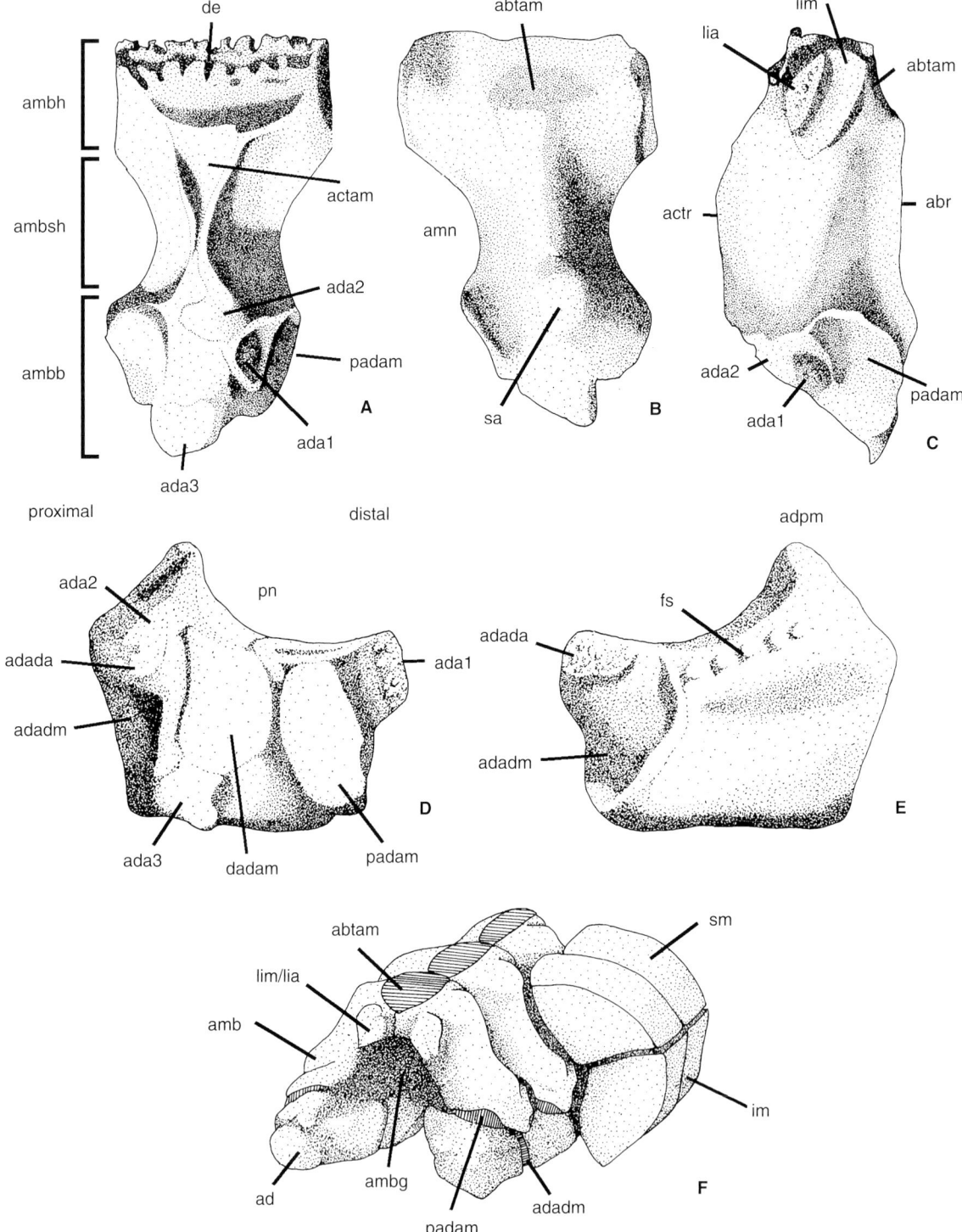

TEXT-FIG. 4. Ambulacral column construction and nomenclature in neoasteroids, illustrated by *Goniopecten demonstrans*, after Gale (1987). A–C, ambulacral ossicle in actinal, abactinal and proximal views. D, E, adambulacral ossicles in abactinal and actinal views. F, ambulacral groove construction. See Table 2 for abbreviations.

The ambulacral base (*ambb*) is the most complex part of the ossicle, and because of its actinal side it contacts two adambulacral ossicles with articulation surfaces and muscle attachment sites (Text-fig. 8A–C). The construction and nomenclature of this region were described in detail by Blake (1973) for paxillosid asteroids, and usage

TEXT-FIG. 5. Interambulacral head articulations and ambulacral angle in abactinal view in: A, *Calliasterella mira*. B, *Astropecten irregularis*. C, *Protoreaster nodosus*. D, *Crossaster papposus*. See Table 2 for abbreviations.

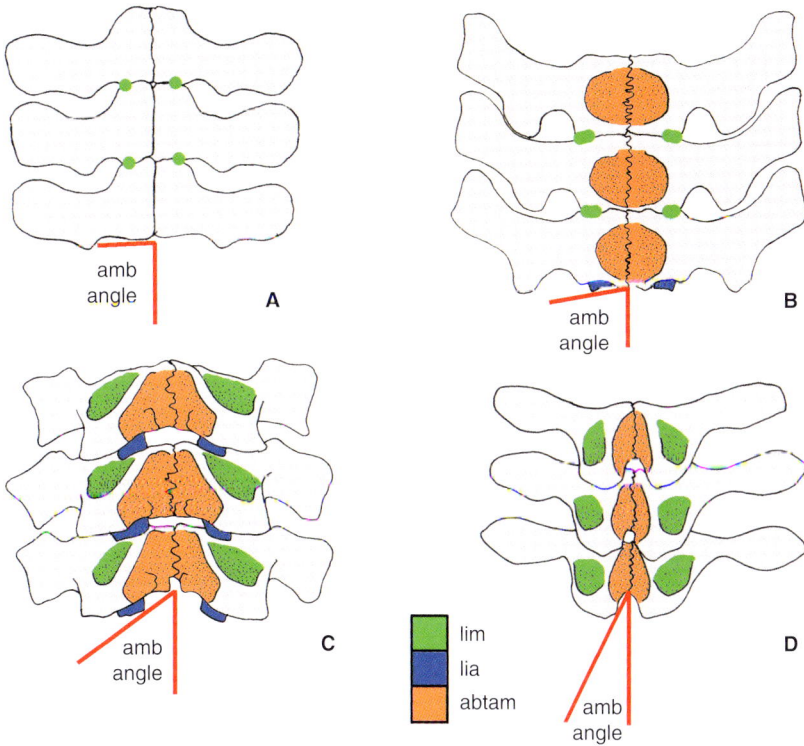

here is modified from his description (Tables 3, 4). In particular, the practice of having separate nomenclature for opposing halves of articulating facets is abandoned. The ambulacral base (Text-fig. 8A–C) comprises a transverse articular ridge on its actinal surface and proximal and distal wing-like processes on which proximal (*padam*) and distal (*dadam*) adambulacral–ambulacral muscles insert (M1 and M2 of Blake 1973; Text-fig. 8D–F herein). In *Calliasterella mira* (Text-fig. 6D–G), *Luidia ciliaris* (Pl. 6, fig. 3), *Astropecten irregularis* (Pl. 6, fig. 11), *Ctenodiscus crispatus* (Pl. 6, fig. 6), *Styracaster chuni*, *Radiaster tizardi*, *Benthopecten simplex* (Pl. 6, fig. 7) and *Archaster lorioli* (Pl. 7, fig. 2), the wings are elongated and strongly asymmetrical, whereas in all other taxa they are short, inconspicuous and nearly symmetrical.

Articulation surfaces, which contact two adjacent adambulacrals on the actinal surface of the ambulacral base, are arranged on the transverse ridge (Text-fig. 8D). In *Calliasterella mira* (Text-figs 6D, 7B–D), *Luidia ciliaris* (Text-fig. 7E, F; Pl. 6, fig. 3), *Radiaster tizardi* and *Astropecten irregularis* (Pl. 6, fig. 11), this comprises a proximal, transverse, strip-like articulation surface (*ada1*) that contacts the distal adambulacral and a distal knob-like, adradial articulation (*ada2*) that contacts a corresponding facet on the proximal adambulacral (Pl. 6). The simple transverse, ridge-like form of *ada1* is found in *Calliasterella mira*, all paxillosids and *Odontaster validus* (Pl. 9, figs 5–6); in all other taxa, separate adradial and abradial contacts (*ada1a, ada1b*) are present, which articulate with

surfaces on the distal transverse ridge of the adambulacral base. In the taxa listed above, a distal abradial surface (*ada3*) on the proximal abradial adambulacral articulates with a distal, abradial facet on the actinal surface of the adjacent adambulacral of the column (Blake 1973; Text-fig. 7A, E, F herein). In all other taxa, the *ada3* on the proximal adambulacral contacts the adradial side of the transverse ridge of the ambulacral base (e.g. Text-fig. 8A, B, D–F). Thus, all neoasteroids, except paxillosids, have four more or less symmetrical articulation surfaces between ambulacrals and adambulacrals (Text-fig. 8).

Superambulacrals, which act as transverse struts between the abactinal surface of each ambulacral base and the inner surface of the inferomarginals (Heddle 1967), are found in *Luidia ciliaris*, *Astropecten irregularis* (Pl. 5, fig. 1), *Ctenodiscus crispatus* (Pl. 5, fig. 2) and *Radiaster tizardi*. Homologous tiny rounded ossicles in contact with the adambulacrals are present immediately abradial to the ambulacral base in *Nardoa variolata* and *Mediaster aequalis* (noted by Fisher 1911).

Adambulacral ossicles

The adambulacral (= exambulacralia of Schäfer 1962, fig. 57) ossicles form two proximally imbricating series of plates extending from the distal surface of the oral ossicle to the terminal, fringing the ambulacral groove. They are joined to each other by interadambulacral muscles and

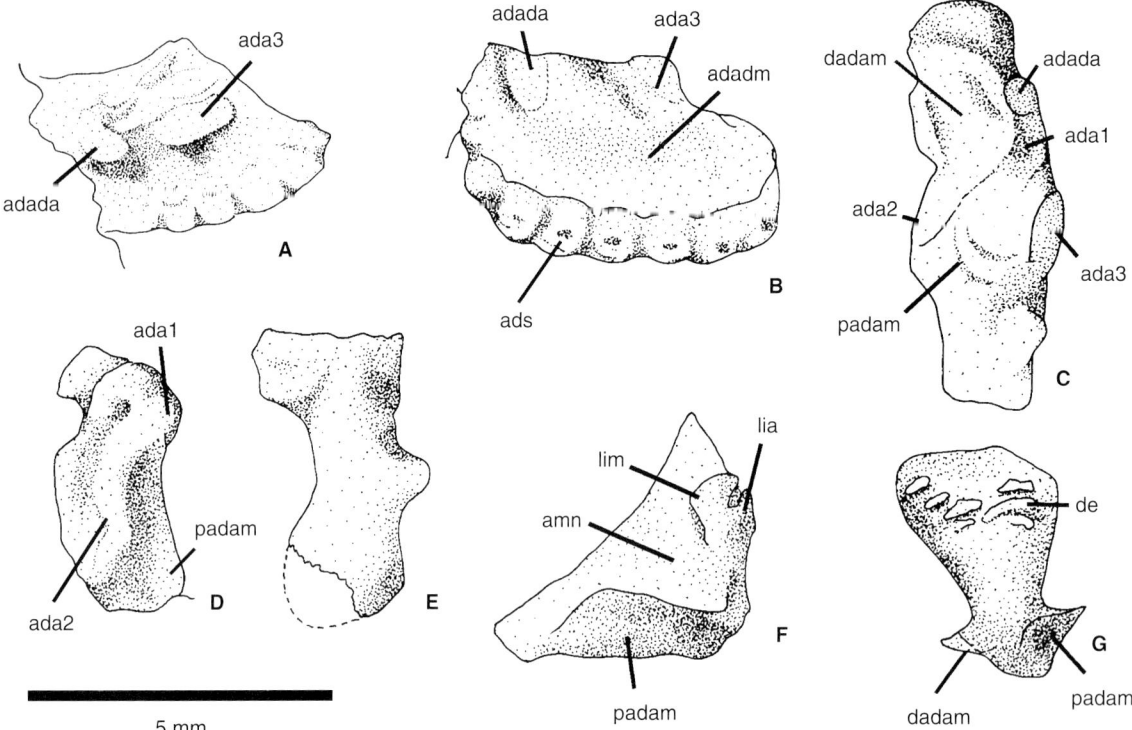

TEXT-FIG. 6. Ambulacral and adambulacral ossicles of *Calliasterella mira*. Specimen figured by Schöndorf (1909), collections of PIN, Moscow. A–C, adambulacrals, in distal (A), proximal (B) and abactinal (C) aspects. D–G, ambulacrals in (D) actinal, (E) abactinal, (F) distal and (G) abradial view. See Table 2 for abbreviations.

articulate by means of interadambulacral articulation surfaces (Text-fig. 8). Tube feet pass through depressions on the adradial surface of the adambulacrals (Text-fig. 8). A free actinal surface carries the adambulacral and subadambulacral spines. The adambulacrals alternate with the ambulacral ossicles, with which they make contact on specialized abactinal articulation surfaces and are joined by proximal and distal ambulacral–adambulacral muscles, which leave clear attachment sites (*padam*, *dadam*; see above).

Although homologies of all adambulacral articulation surfaces and muscle insertion sites found in neoasteroids are identifiable in *Calliasterella mira*, important differences in proportion and topology exist between neoasteroids and the Carboniferous taxon. In the latter, the adambulacrals comprise an abactinal portion, which imbricates proximally and a vertically oriented actinal part that bears the interadambulacral muscles (Pl. 17, figs 7–8; Text-figs 6A–C, 7B–D). In all neoasteroids, the adambulacrals are rhomboidal in lateral aspect and imbricate proximally (Text-figs 7E, F, 8). The transformation from the condition in *Calliasterella mira* can be made by simple topological shift of landmark structures, involving an oblique actinal–abactinal flattening of the ossicle, such

---

EXPLANATION OF PLATE 6

SEM images of cleaned ambulacral and adambulacral ossicles of extant neoasteroids. Proximal to left of plate, distal to right. Figs 2–3, 5–6 and 7–8 represent articulating pairs of ossicles. Adradial towards top of page.

Figs 1–3, 9. *Luidia ciliaris*, actinal and abactinal view of adambulacral ossicle, actinal view of ambulacral base and enlarged view of distal *ada3* articulation surface seen in Fig. 1, respectively.

Fig. 4. *Styracaster chuni*, abactinal view of adambulacral ossicle.

Figs 5–6. *Ctenodiscus crispatus*, abactinal view of adambulacral and actinal view of ambulacral base, respectively.

Figs 7–8. *Benthopecten simplex*, abactinal view of adambulacral and actinal view of ambulacral base, respectively.

Figs 10–11. *Astropecten irregularis*, abactinal view of adambulacral and actinal view of ambulacral base, respectively.

See Table 2 for abbreviations.

Figs 1–2, 6–7, ×20; Figs 3–4, ×25; Figs 10–11, ×30; Fig. 9, ×75.

PLATE 6

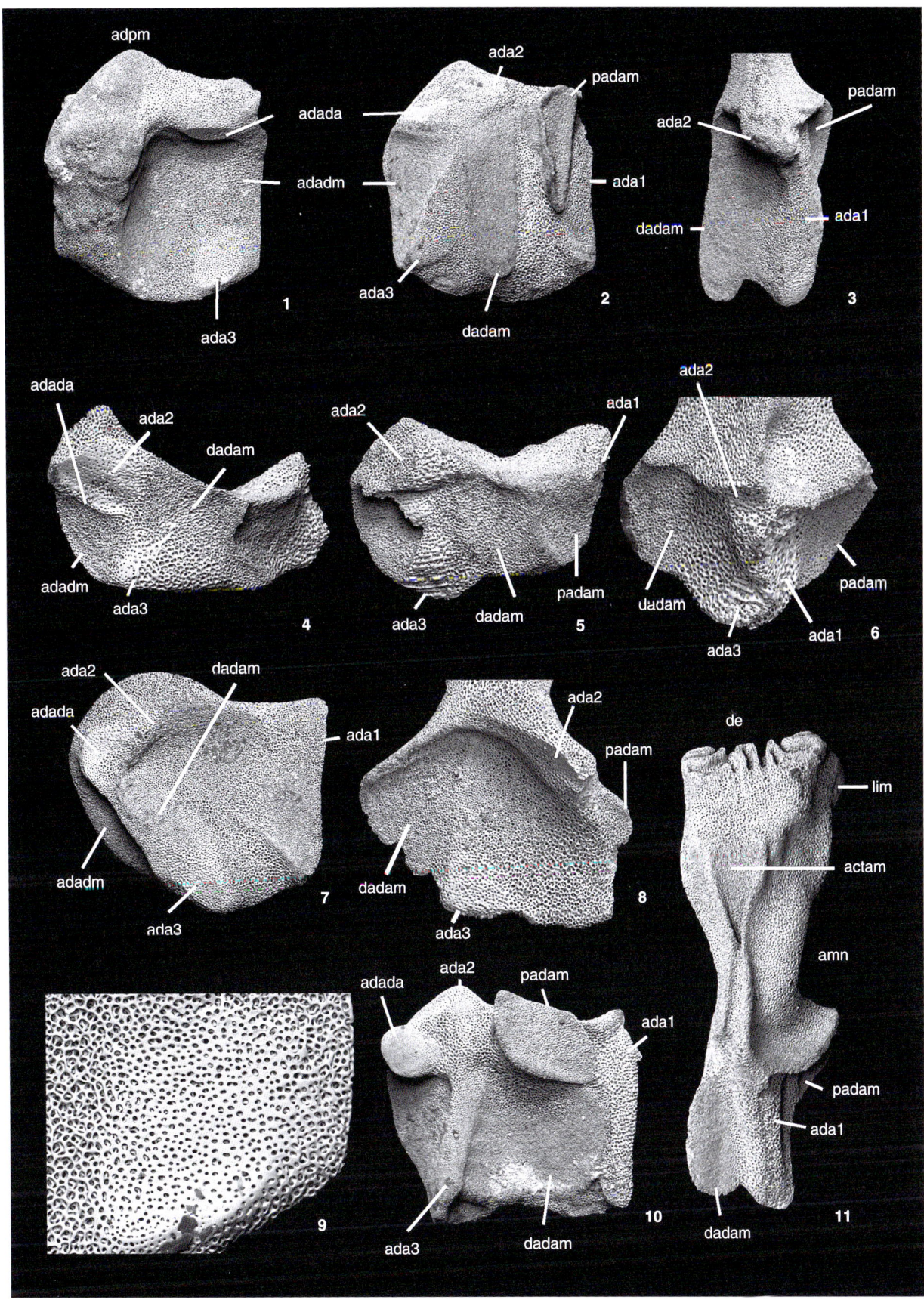

GALE, Ambulacral and ambulacral morphology of Recent asteroids

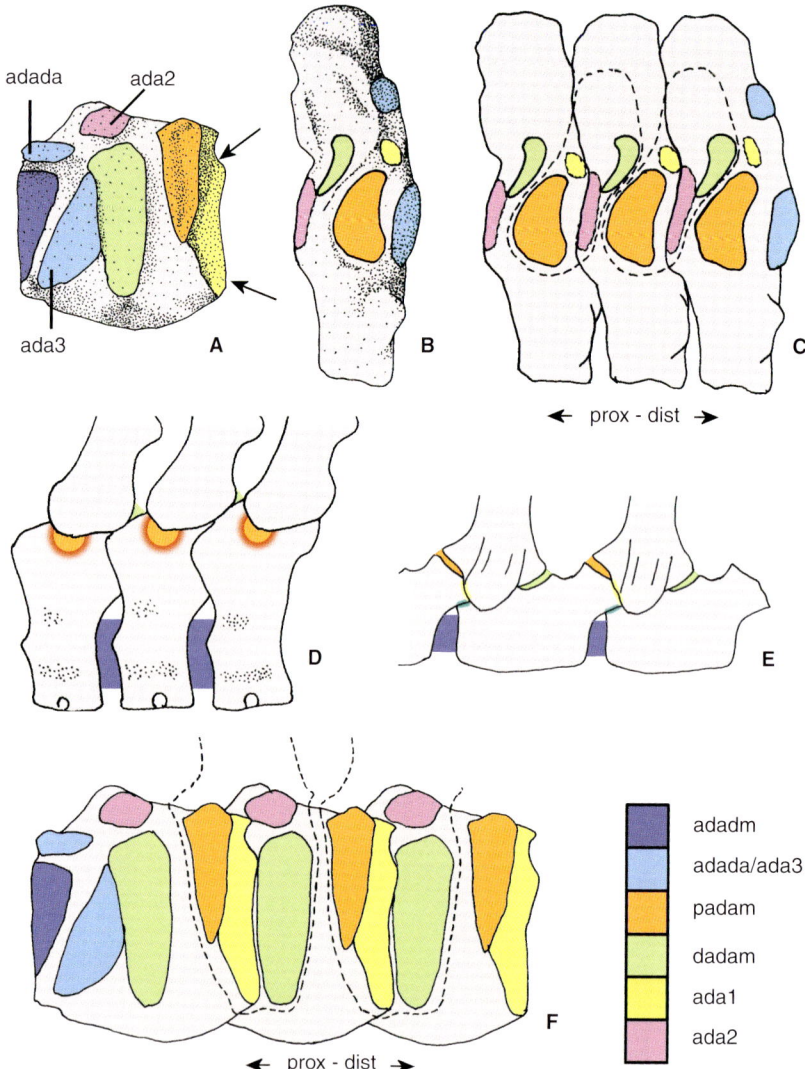

TEXT-FIG. 7. Comparison of ambulacral–adambulacral articulations and muscles in *Calliasterella mira* and *Luidia ciliaris*. Actinal aspects of single (A) and articulated adambulacrals, and (E) abradial view of ambulacral–adambulacral articulation of *Luidia ciliaris*; actinal aspects of single (B) and articulating (C) adambulacrals of *Calliasterella mira*. D, reconstruction of adambulacral–ambulacral articulation in lateral view in *Calliasterella mira*. F, abactinal view of articulating adambulacrals in *L. ciliaris*, position of ambulacrals outlined by dashed lines. Compare with Text-figure 8D, E. Not to scale. See Table 2 for abbreviations.

EXPLANATION OF PLATE 7

SEM images of cleaned ambulacral and adambulacral ossicles of extant neoasteroids. Proximal to left of plate, distal to right.

Figs 1–2, 4–5 and 8–9 represent articulating pairs of ossicles. Adradial towards top of page.

Figs 1–2. *Archaster lorioli*, abactinal view of adambulacral and actinal view of ambulacral base, respectively.

Figs 3–5. *Nardoa variolata*, oblique proximal view of adambulacral, abactinal view of adambulacral and actinal view of ambulacral base, respectively.

Fig. 6. *Asteropsis carinifera*, abactinal view of adambulacral.

Fig. 7. *Mediaster aequalis*, abactinal view of adambulacral.

Figs 8–11. *Echinaster purpureus*, abactinal view of adambulacral, oblique actinal view of ambulacral base, proximal oblique and abactinal views of ambulacral, respectively.

See Table 2 for abbreviations.

Fig. 1, ×12; Figs 2, 10–11, ×20; all others, ×25.

PLATE 7

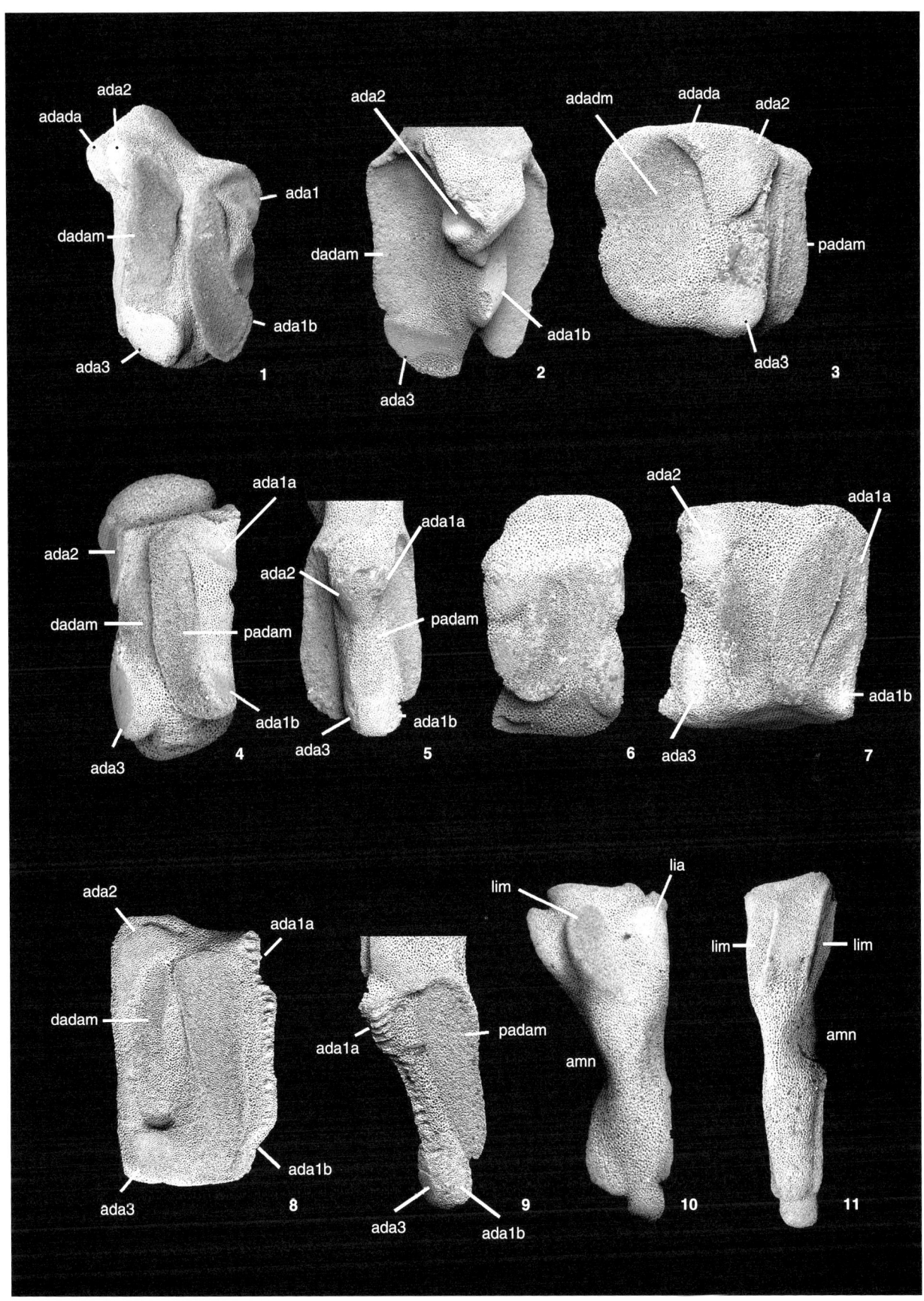

GALE, Ambulacral and ambulacral morphology of Recent asteroids

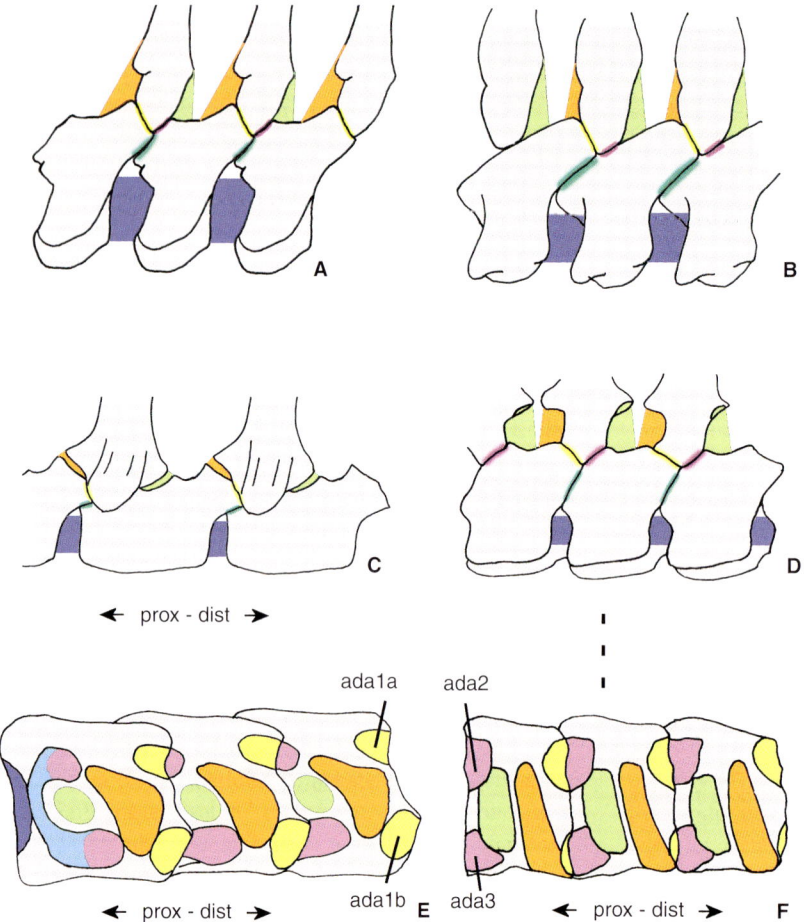

TEXT-FIG. 8. Adambulacral–ambulacral articulations and muscles in neoasteroids: A, *Odontaster validus.* B, *Crossaster papposus.* C, *Luidia ciliaris.* D, F, *Nardoa variolata.* E, *Porania pulvillus.* A–D, lateral aspect. E, F, abactinal view of articulated adambulacrals. Note that proximal and distal articulation structures between ambs and adambs are approximately symmetrical in all taxa except *Luidia ciliaris.* Coloured homologies as in Text-figure 7. See Table 2 for abbreviations.

that the distal surface of the adambulacral is not visible from an actinal view. In *Calliasterella mira*, the adambulacrals are very broad in proportion to the ambulacrals, whereas in neoasteroids the ambulacrals are invariably broader than the adambulacrals (Text-fig. 1).

A distal adambulacral–ambulacral articulation surface (*ada1* = Pa1 of Blake 1973) is invariably present between the distal part of the adambulacral and the proximal transverse ridge of the ambulacral base. However, the nature of this articulation is highly variable within neoasteroids. In *Astropecten irregularis, Radiaster tizardi* and *Luidia ciliaris* (compare Blake 1973), it comprises a triangular or rectangular area of sculptured stereom, usually with a central groove, which forms a flat contact with a corresponding surface on the proximal ambulacral base and extends across the width of the ambulacral base (Pl. 6, figs 2, 10; Text-figs 7A, E, F, 10A–C). A similar construction is present in *Benthopecten simplex* (Pl. 6, fig. 7; Text-fig. 10D) and *Odontaster validus* (Text-fig. 11L). In *Ctenodiscus crispatus* and *Styracaster chuni*, a small rounded *ada1*, made of rugose stereom, articulates with a pitted depression in the adradial proximal region of the ambulacral base (Turner

and Dearborn 1972; Pl. 6, figs 4–6; Text-figs 10E, F herein).

In all neoasteroids, except paxillosids and *Odontaster validus*, *ada1* is separated into two discrete surfaces (abradial, *ada1b*; adradial, *ada1a*) constructed of smooth imperforate stereom (Pls 7–9; Text-fig. 8). These are concave and articulate with two convex, smooth bosses on the proximal part of the corresponding ambulacral base (e.g. Pl. 7, figs 4–5; Text-fig. 8E, F). There is significant variation between taxa in the shape, relative proportions and precise positions of these two articular surfaces (see Pls 7–9; Text-figs 10, 11).

A proximal adambulacral–ambulacral articulation (*ada2* = Pa2 of Blake 1973) is also present in *Calliasterella mira* and all neoasteroids. This is a transversely broad and short, poorly defined, abradial depression in *Calliasterella mira* (Text-figs 6C, 7B, C). A single small, rather inconspicuous, boss-like, adradial articulation surface is present on the proximal adambulacral in *Luidia ciliaris* (Pl. 6, fig. 2; Text-fig. 7F), *Radiaster tizardi* (Text-fig. 10B) and *Astropecten irregularis* (Pl. 6, fig. 10; Text-fig. 10C), which contacts an equivalent surface on the distal adradial ambulacral base (Pl. 6, figs 2–3, 10–11).

In most other neoasteroids, *ada2* comprises a flat or concave area of smooth or ridged imperforate stereom, which articulates with the ambulacral base (e.g. *N. variolata*; Pl. 7, fig. 4; Text-figs 8D, F, 10G). In some taxa, *ada2* and *ada3* have fused to form a single articular surface (see below).

In *Calliasterella mira*, paired interadambulacral articulation surfaces are conspicuous and well-developed oval regions of smooth stereom present on proximal and distal sides of successive adambulacrals. In abactinal aspect (Text-fig. 7B, C), only the distal pair is visible. These are identified as homologous with *ada3* and *adada* in neoasteroids (Text-figs 7B, C, 11A–C). The proximal surfaces are slightly concave, the distal ones gently convex. In *Luidia ciliaris* (Pl. 6, figs 1–2) and *Radiaster tizardi* (Text-fig. 10B), four discrete regions of imperforate stereom are homologous with the interadambulacral articular structures in *Calliasterella mira*. However, the actinal–abactinal slanting of the adambulacrals, which allows the proximal imbrication of the ossicles, has brought about topological changes in the positions of articular structures in neoasteroids. The distal articulation surfaces in *Luidia ciliaris* and *Radiaster tizardi* are present on the actinal face of the adambulacrals, and *adada* and *ada3* have migrated distal relative to the interadambulacral muscle *adadm* (Pl. 6, figs 1–2). A similar arrangement is seen in *Astropecten irregularis* (Pl. 6, fig. 10; Text-fig. 10C) and *Benthopecten simplex* (Pl. 6, fig. 7; Text-fig. 10D), but the distal abradial articulation surface *ada3* is no longer present. In all other neoasteroids, the distal interadambulacral articulation surface *ada3* comprises a diffuse transverse region of labyrinthic stereom on the distal margin of the actinal face of the adambulacral (e.g. Pl. 9, figs 1–2).

The proximal *adada* articulation structure is a sharply defined region of smooth stereom in *Archaster lorioli* (Pl. 7, fig. 1; Text-fig. 10H) and *Asteropsis carinifera* (Pl. 7, fig. 6, Text-fig. 10L), and present but poorly defined in *Mediaster aequalis* (Pl. 7, fig. 7), *Nardoa variolata* (Pl. 7, figs 3–4), *Protoreaster nodosus* (Text-fig. 10I) and *Acanthaster planci* (Text-fig. 10J). The proximal *adada* surface is also present in *Zoroaster fulgens* (Pl. 8, fig. 5; Text-fig. 11E), *Asterias rubens* (Pl. 8, fig. 10) and *Heliaster helianthus*. It is absent in *Crossaster papposus* (Pl. 9, fig. 1; Text-fig. 11K), *Asterina gibbosa* (Pl. 9, fig. 8; Text-fig. 11J), *Odontaster validus* (Pl. 9, fig. 5; Text-fig. 11L), *Porania pulvillus* (Text-fig. 11I), *Remaster gourdoni* (Pl. 9, figs 9–10; Text-fig. 11G) and *Pteraster pulvillus* (Pl. 9, fig. 12; Text-fig. 11H).

Ada3 exists as an exclusively interadambulacral articulation in *Calliasterella mira* (Text-fig. 6A–C), *Luidia ciliaris* (Text-fig. 7A, F) and *Astropecten irregularis* (Pl. 6, fig. 10) (see also Blake 1973). In all other neoasteroid taxa, *ada3* articulates with both the proximal adambulacral and an articular structure on the abradial proximal side of the

ambulacral base (Text-fig. 8A, B, D–F). In *Nardoa variolata* (Pl. 7, fig. 4; Text-figs 8F, 10G), *Archaster lorioli* (Text-fig. 10H; Pl. 7, figs 1–2), *Protoreaster nodosus* (Text-fig. 10I), *Asteropsis carinifera* (Pl. 7, fig. 6) articular surfaces *ada2* and *ada3* are symmetrical and together with the abradial and adradial parts of *ada1a* and *ada1b* form two shallow, cup-shaped depressions that articulate with the convex articular surfaces on the adradial and abradial sides of the ambulacral base (e.g. *Nardoa variolata*; Text-fig. 8D, F; compare with Pl. 7, figs 4–5). Thus, the adambulacral–ambulacral articulation has a 'double-roller' construction, with an interadambulacral articulation adjacent to and contiguous with the ambulacral–adambulacral surface (see Text-fig. 8D, F) in many neoasteroids.

However, in certain taxa, *ada3* and *ada3* are confluent or fused to form a single, transverse crescentic (*Crossaster papposus*; Text-fig. 11K; *Porania pulvillus*; Text-figs 8E, 11I) or hourglass-shaped (*Asterina gibbosa*; Text-fig. 11J; *Odontaster validus*; Text-fig. 11L), flat or slightly concave articular surface. The distal part of this facet contacts two convex articular surfaces on the distal ambulacral base, and its proximal part acts as an interadambulacral articulation surface, contacting smooth perforate stereom on the actinal distal part of the adjacent adambulacral (e.g. *Porania pulvillus*; Text-fig. 8E; *Crossaster papposus*; Text-fig. 11K; Pl. 9, figs 1–2).

The most dramatically modified adambulacrals are found in the korethrasterid *Remaster gourdoni* and the pterasterid *Pteraster pulvillus*. Fisher (1940, fig. C, 5) noted that fusion (ankylosis) of inferomarginal and adambulacral ossicles had occurred in the korethrasterid *Peribolaster folliculatus*, an observation supported by Downey (*in* Clark and Downey 1992). However, there is no evidence that fusion has taken place and the ossicles appear to be single calcite crystals. The broad, short abradial process is herein called the adambulacral extension (*adex*). In *Remaster gourdoni*, this is a transversely broad, short, curved ossicle (Pl. 9, figs 9–10; Text-figs 11G, 12A). The insertion sites of ambulacral–adambulacral muscles *padam* and *dadam*, the distal double articulation facets *ada1*, the fused *ada2*, *ada3* articulation facet and the insertion site of the interadambulacral muscle can all be identified on the adradial part of the adambulacrals. The adambulacrals are distinctly flattened with a strong imbrication. A bifid attachment site for a large spine (*als*) is present on the lateral actinal margin of the ossicle. A proximal and a distal interadambulacral extension muscle (*adexm*) and a centrally placed articulation facet (*adexa*) are present on the lateral portion of the adambulacral extension. The adambulacrals of *Pteraster pulvillus* (Pl. 9, figs 11–12; Text-figs 11H, 12B) are more flattened and more strongly imbricate than those of *Remaster gourdoni* (Pl. 9, figs 9–10). The large, lateral-webbed adambulacral spines of *Pteraster*, which support the actinolateral

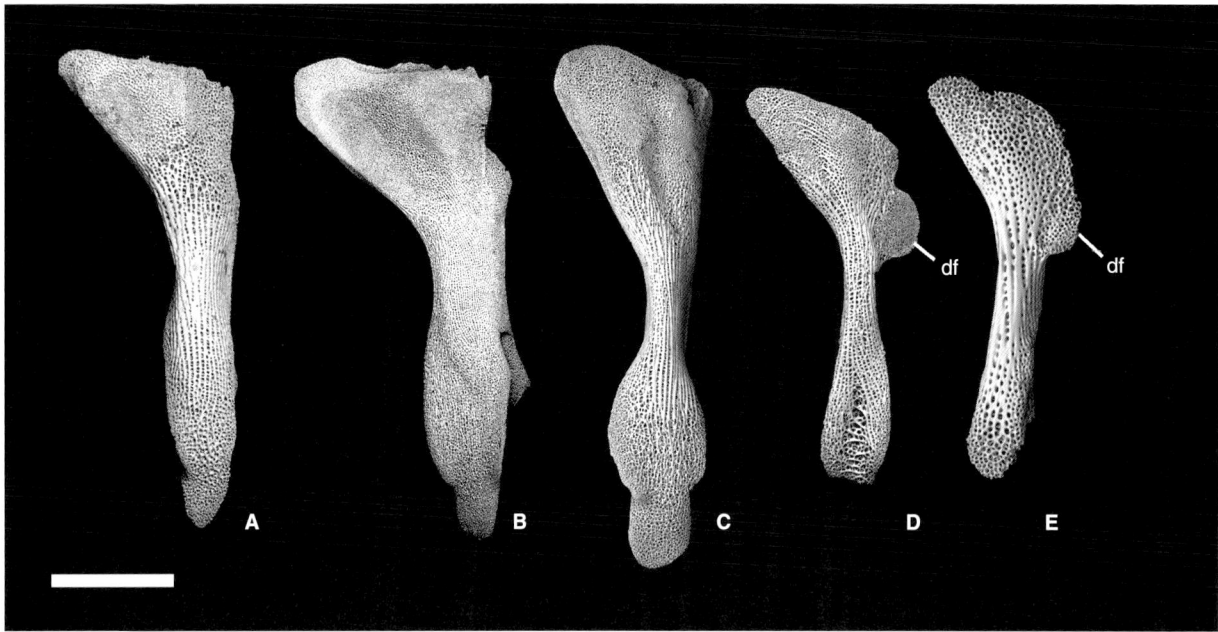

TEXT-FIG. 9. Ambulacral ossicles in abactinal view of present-day neoasteroids; from left to right (A–E) are *Asterina gibbosa*, *Crossaster papposus*, *Odontaster validus*, *Pteraster pulvillus* and *Remaster gourdoni*. Note the transversely elongated glassy trabeculae that make up the ambulacral shafts, the enlarged proximal flange on the ambulacral heads and the distal flange in *Remaster* and *Pteraster*. Scale bar represents 1 mm. Proximal left, distal right. See Table 2 for abbreviation.

membrane, are homologous with the lateral actinal spine in *Remaster gourdoni* (Text-fig. 12).

Mouth frame

The construction and homologies of the mouth region and its constituent ossicles have received rather little attention in asteroids since the detailed descriptions of Viguier (1879), with the exception of Turner and Dearborn's (1972; see Table 5) account of *Ctenodiscus crispatus*. In the outgroup *Calliasterella mira* and all neoasteroids, the mouth frame is made up of 10 oral ossicles, 10 circumorals

(Pl. 1, fig. 2) and five unpaired interradial ossicles, traditionally called axilliaries in Palaeozoic asteroids (Shackleton 2005) and odontophores in neoasteroids (Viguier 1879). In most Palaeozoic asteroids, the axillary articulates with the marginals and the oral ossicles and has a relatively large external face (Shackleton 2005). The homologous odontophore in neoasteroids is largely internal and does not articulate with the marginals, and a small external face is present only in some paxillosids (e.g. Turner and Dearborn 1972).

The pairs of oral ossicles connect with each other interradially by means of articulation structures and two muscles (actinal and abactinal interradial interoral muscles,

---

PLATE 8

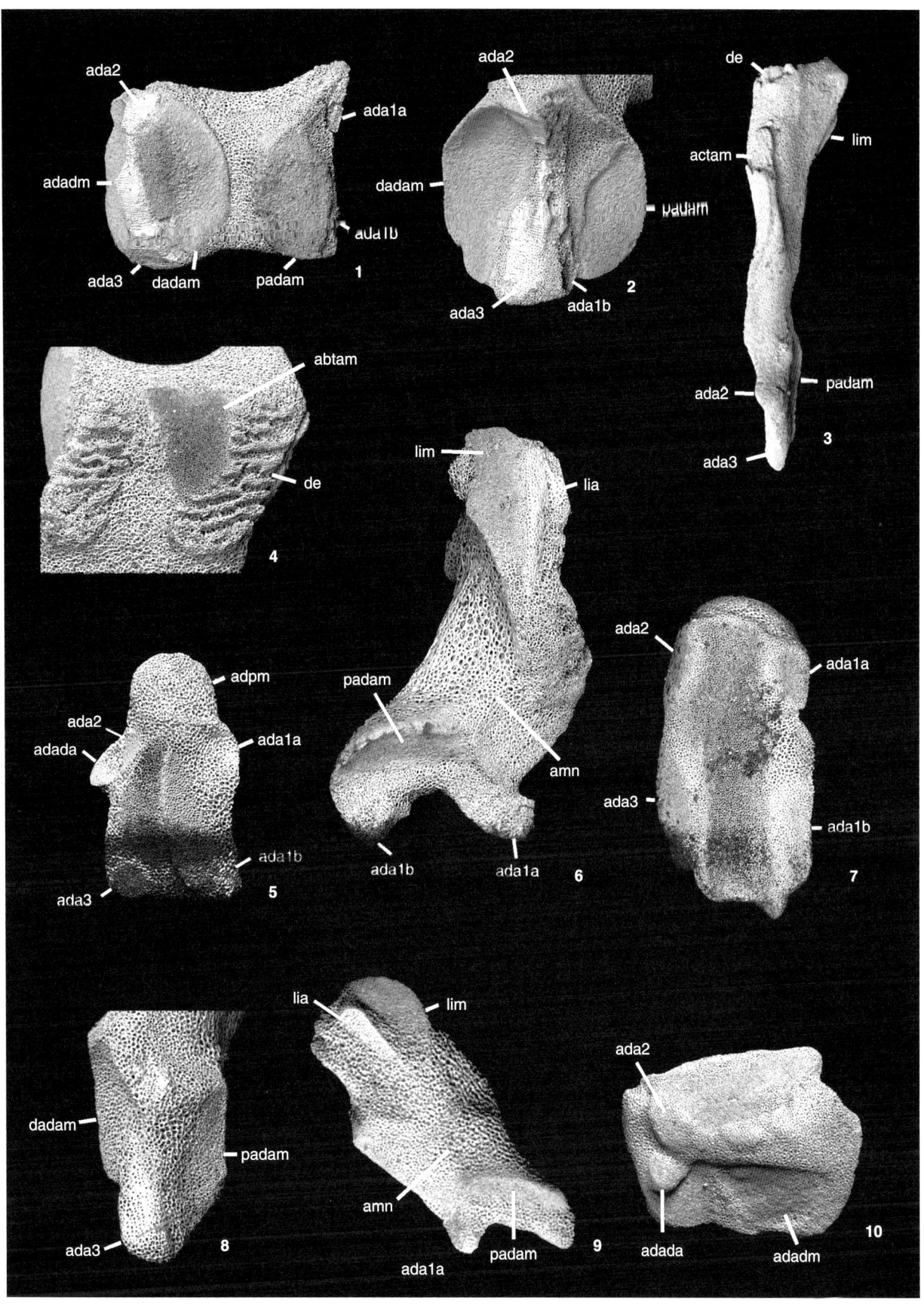

GALE, Ambulacral and ambulacral morphology of Recent asteroids

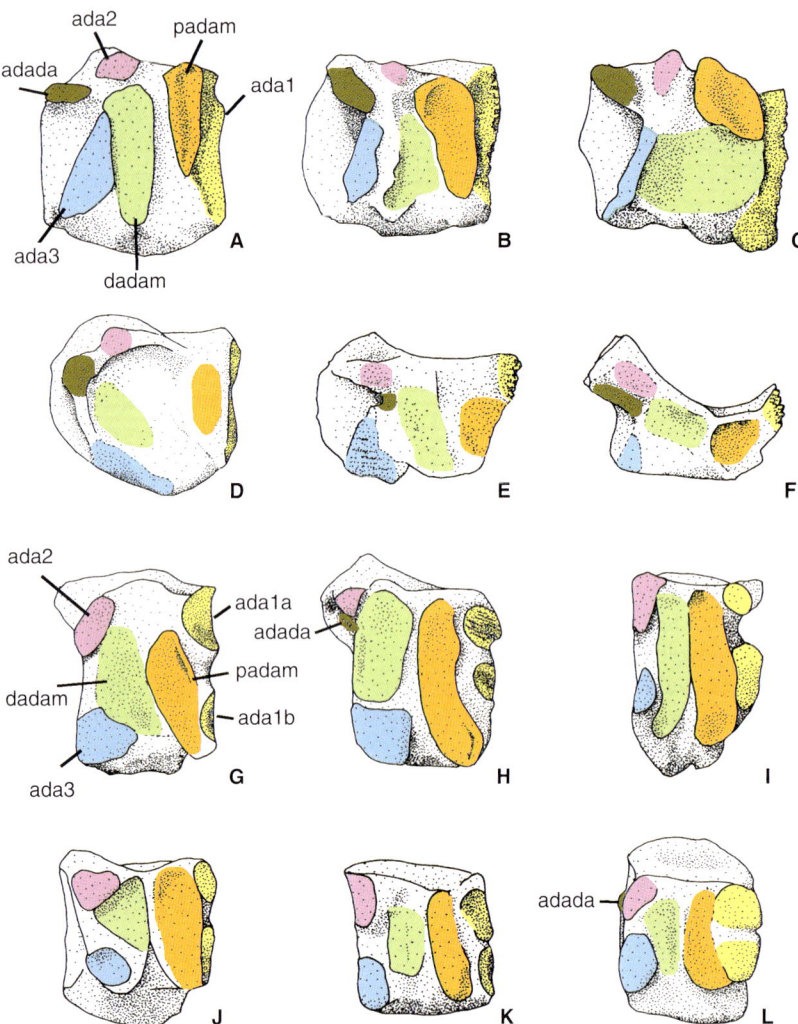

TEXT-FIG. 10. Drawings of adambulacral ossicles in abactinal aspect to show homologies of articulations and muscle insertion sites. A, *Luidia ciliaris*. B, *Radiaster tizardi*. C, *Astropecten irregularis*. D, *Benthopecten simplex*. E, *Ctenodiscus crispatus*. F, *Styracaster chuni*. G, *Nardoa variolata*. H, *Archaster lorioli*. I, *Protoreaster nodosus*. J, *Acanthaster planci*. K, *Mediaster aequalis*. L, *Asteropsis carinifera*. Proximal to left, distal to right. Amb groove to the top of the page. Not to scale. See Table 2 for abbreviations.

aciim, abiim; the 'interdentaire' and 'adducteur' of Viguier 1879, fig. A), and a single triangular, rectangular or oval odontophore lies between each pair (Pl. 1, fig. 2; Text-fig. 13). The odontophore articulates with the inner, distal surface of the orals by means of one to three paired articulation surfaces composed of smooth perforate or imperforate stereom (Pls 13–15), and a vertical oral-odontophore muscle (*odom*) runs between the two (Text-fig. 13C). The insertion sites of this muscle are clearly visible as areas of retiform or galleried stereom on both oral ossicles and actinal surface of the odontophore (Pl. 13, figs 1, 3, 6). The actinal surface of the oral ossicle bears oral and suboral spines on articular structures (*osp, sos*). Abactinally, each oral ossicle articulates with a circumoral

---

EXPLANATION OF PLATE 9

SEM images of cleaned ambulacral and adambulacral ossicles of extant neoasteroids. Proximal to left of plate, distal to right. Figs 1–3, 4–5 and 7–12 represent articulating pairs of ossicles. Adradial towards top of page.

Figs 1–3. *Crossaster papposus*, abactinal view of adambulacral, actinal view of adambulacral and actinal view of ambulacral base, respectively.

Figs 4, 8. *Asterina gibbosa*, actinal view of ambulacral base and abactinal view of adambulacral, respectively.

Figs 5–6. *Odontaster validus*, abactinal view of adambulacral and actinal view of ambulacral base, respectively.

Figs 7, 11–12. *Pteraster pulvillus*, actinal view of ambulacral base, actinal and abactinal view of adambulacral, respectively.

Figs 9–10. *Remaster gourdoni*, actinal and abactinal view of adambulacral, respectively.

See Table 2 for abbreviations.

Figs 1–2, 5–6, ×20; Figs 3–4, ×30; Figs 7–8, 11–12, ×40; Figs 9–10, ×110.

PLATE 9

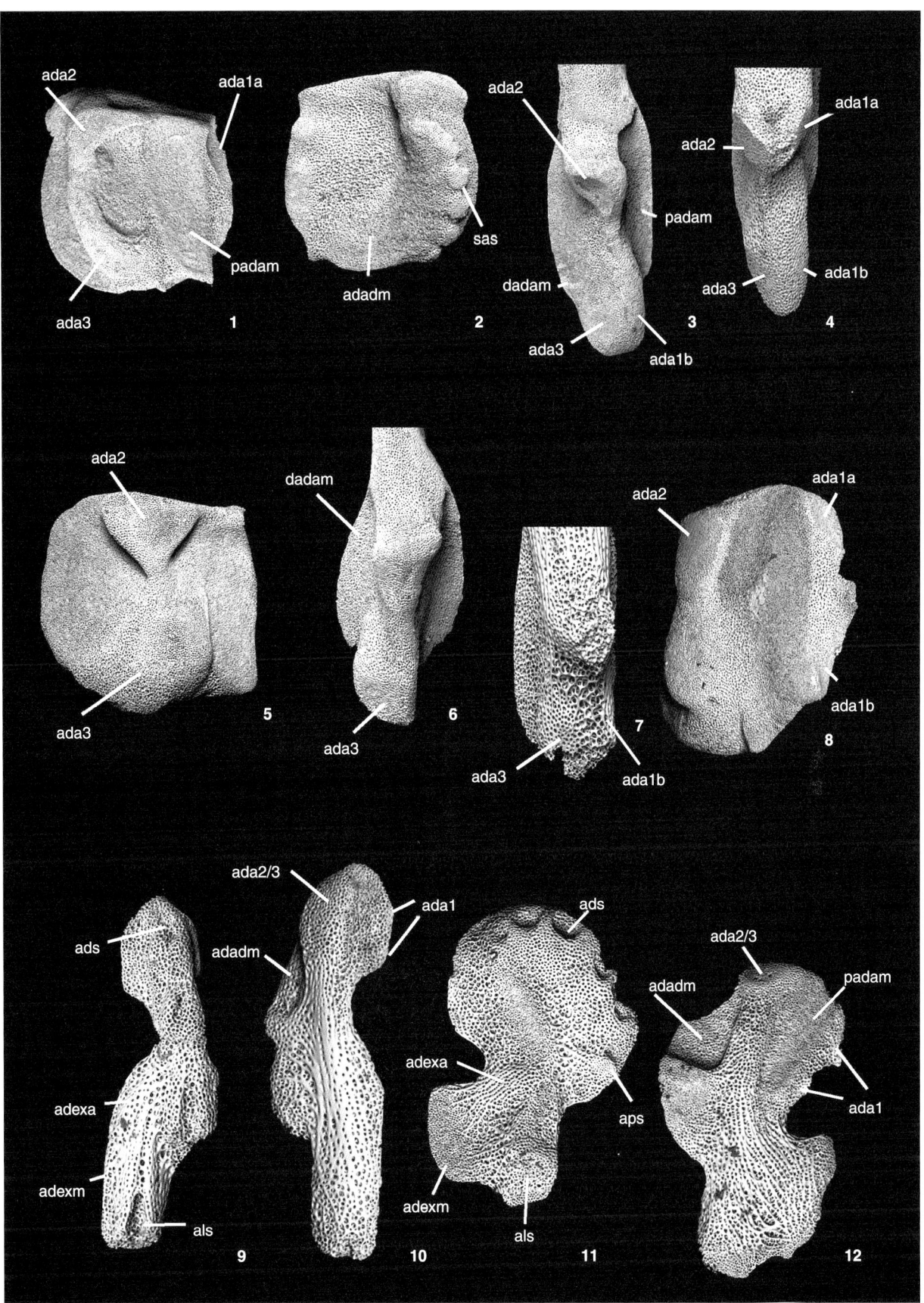

GALE, Ambulacral and ambulacral morphology of Recent asteroids

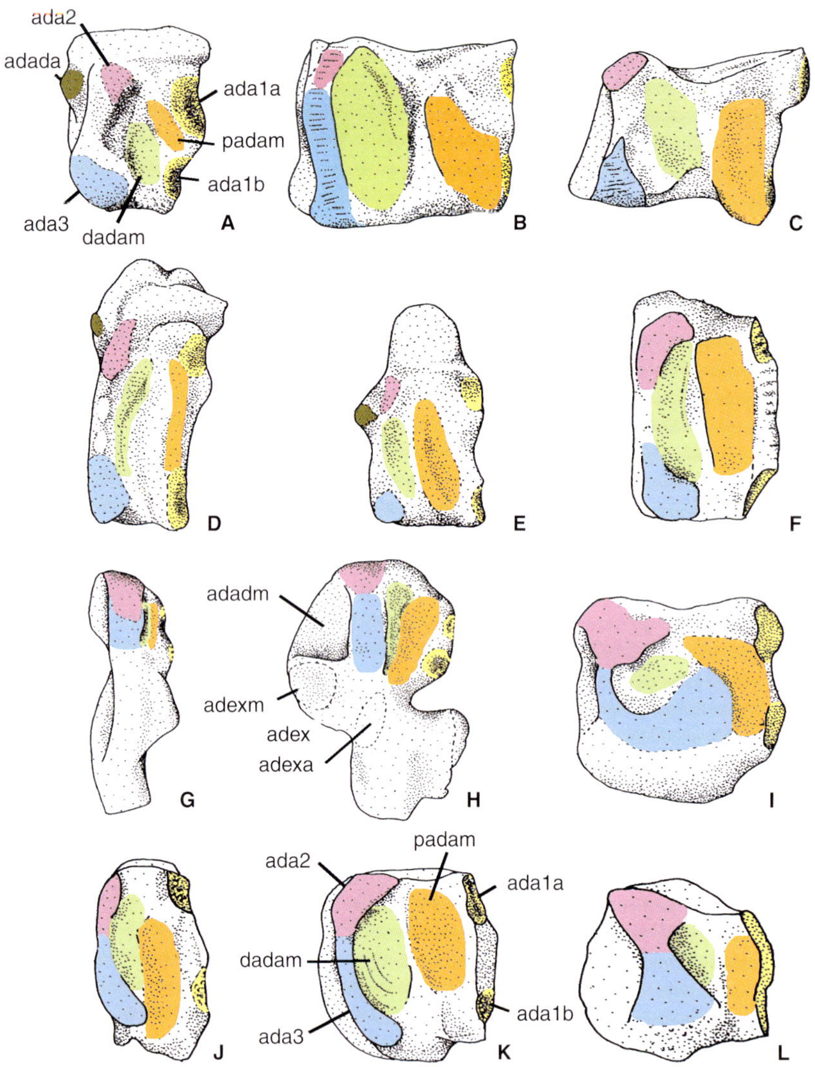

TEXT-FIG. 11. Drawings of adambulacral ossicles in abactinal aspect to show homologies of articulations and muscle insertion sites. A, *Heliaster helianthus*. B, *Brisinga costata*. C, *Freyella elegans*. D, *Asterias rubens*. E, *Zoroaster fulgens*. F, *Echinaster purpureus*. G, *Remaster gourdoni*. H, *Pteraster pulvillus*. I, *Porania pulvillus*. J, *Asterina gibbosa*. K, *Crossaster papposus*. L, *Odontaster validus*. Proximal to left, distal to right. Amb groove to the top of the page. Not to scale. See Table 2 for abbreviations.

by means of a proximal (*pcoa*) and a distal articulation (*dcoa*), and the space between the bars forms the first podial opening (*1st tf*; Text-fig. 13C). The proximal circumoral bar is taller than the distal one and forms part of an enlarged proximal wing called the apophyse (*apo*; Turner and Dearborn 1972), which also carries a groove for the ring vessel of the water vascular system (*rvg*; Text-

fig. 13). The outer (radial) face of the apophyse of each oral ossicle bears a muscle, which inserts on the outer surface of the adjacent pair of orals across the mid-radial line (radial interoral muscles, *riom*; 'abducteur' of Viguier 1879). A conspicuous groove for the ring nerve is present immediately actinal to the apophyse (*rng*; Pl. 13, figs 1, 4, 6, 9). The oral articulates with the first adambulacral by

---

EXPLANATION OF PLATE 10

SEM images of ambulacral heads of cleaned ossicles of extant neoasteroid taxa.

Figs 1–2. *Nardoa variolata*, proximal (1) and distal (2) aspects of ambulacral head.

Figs 3–4. *Crossaster papposus*, proximal (3) and distal (4) aspects of ambulacral head.

Figs 5–6. *Luidia ciliaris*, proximal (5) and distal (6) aspects of ambulacral head. Note the retiform stereom associated with muscle insertion, and the perforate smooth stereom of articular surfaces.

See Table 2 for abbreviations.

All ×100.

PLATE 10

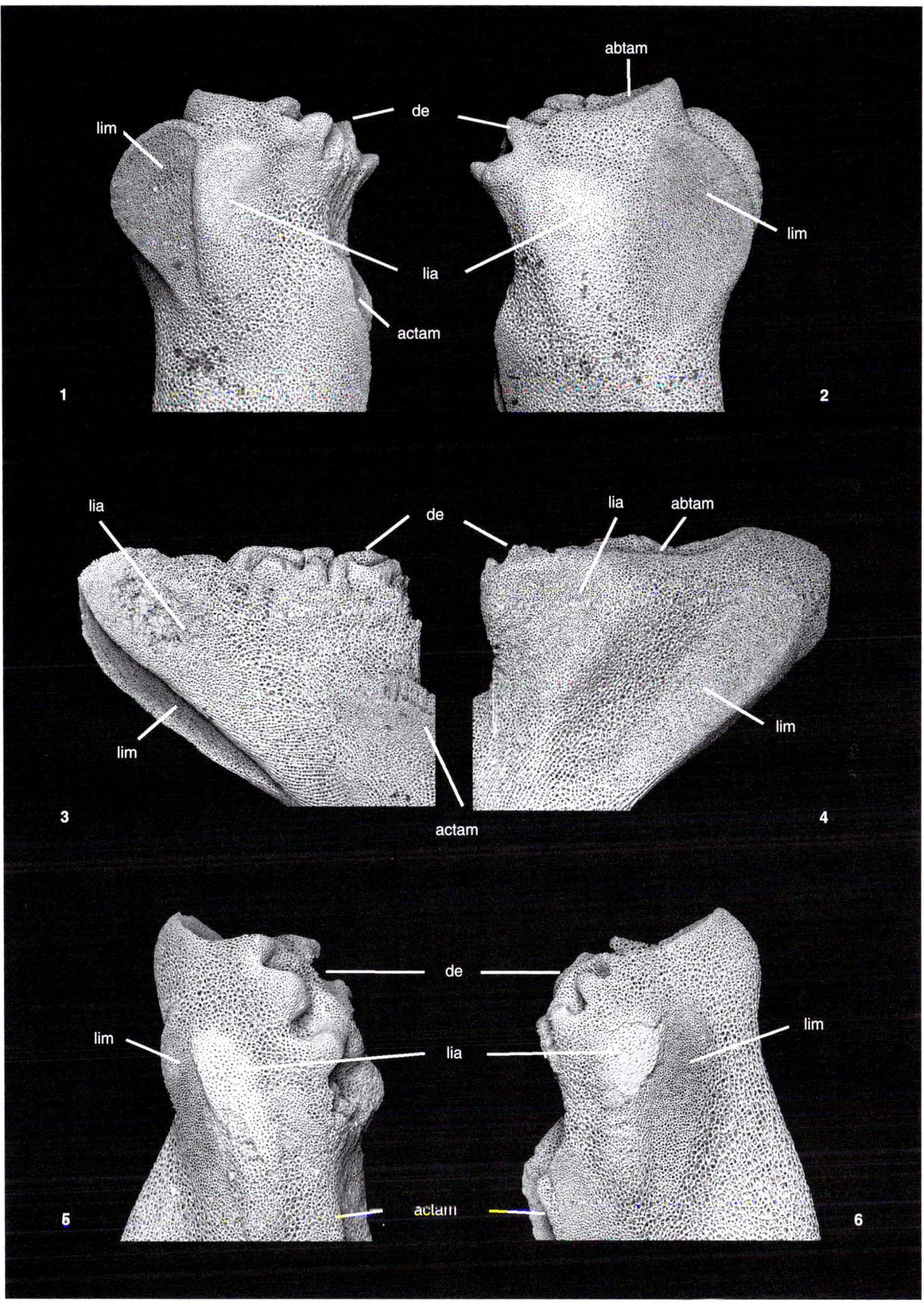

GALE, Morphology of ambulacral heads of Recent asteroids

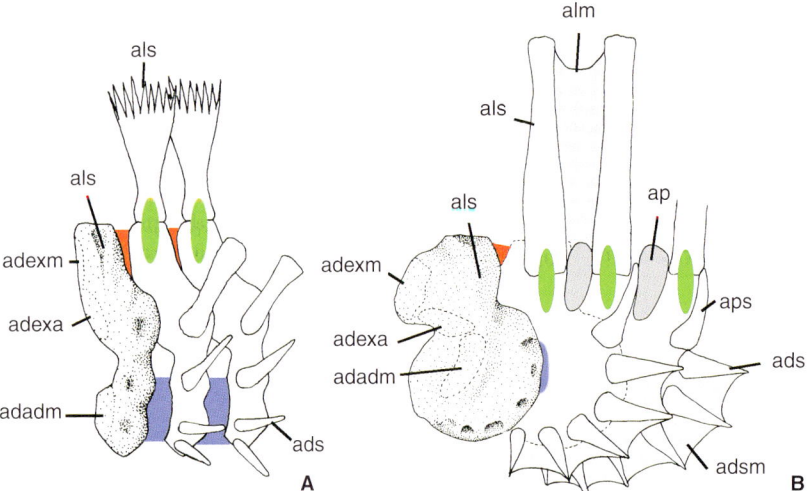

TEXT-FIG. 12. Adambulacrals in *Remaster gourdoni* (A) and *Pteraster pulvillus* (B). Denuded ossicles stippled. The adambulacral extensions (*adex*) have specialized articulation surfaces and muscles (*adexa, adexm*) and carry an enlarged actinolateral spine (*als*). In *Pteraster*, the adambulacral spines are webbed, as are the actinolateral spines. Apertures to the interior of the canopy are guarded by apertural spines. See Table 2 for abbreviations. Proximal left, distal right.

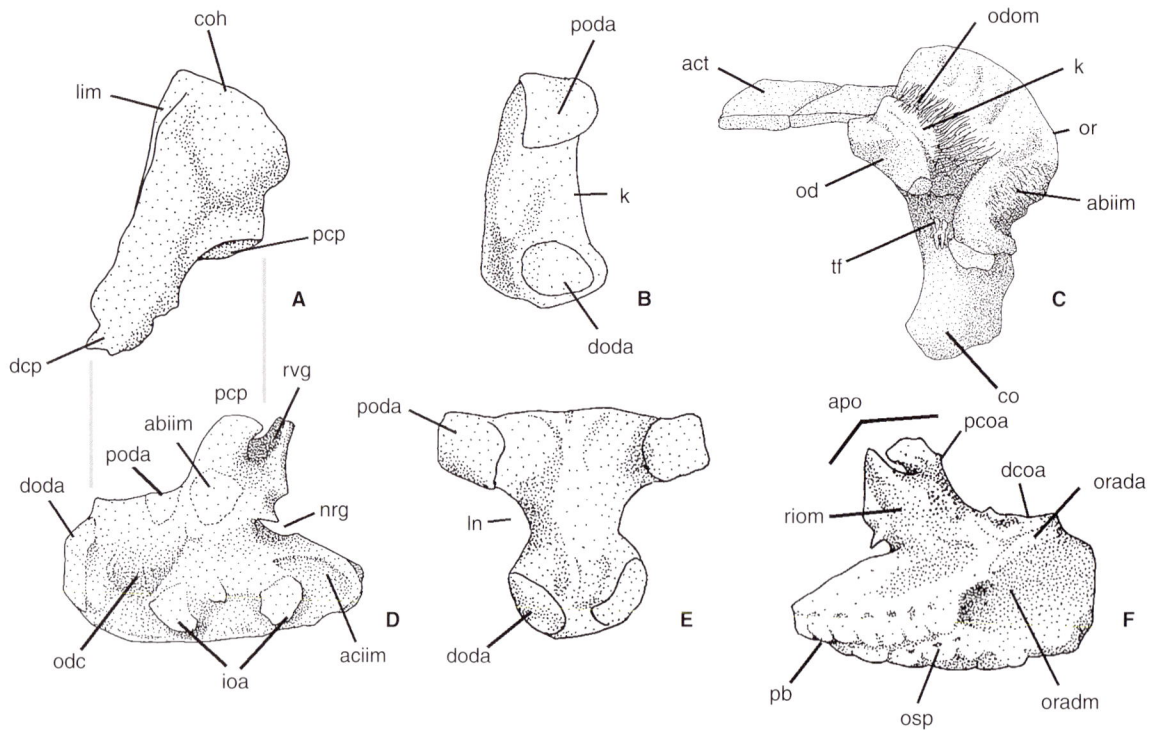

TEXT-FIG. 13. Nomenclature of asteroid mouth frame ossicles illustrated by *Nardoa variolata* (A, B, D–F) and *Goniopecten demonstrans* (C); A, circumoral in lateral aspect. B, E, odontophore in lateral and actinal views, respectively. D, F, oral ossicle in internal (interradial) and external (radial) aspects, respectively. C, dissected mouth frame in interradial aspect to show muscles and arrangement of ossicles. See Table 2 for abbreviations.

means of one or several surfaces (*orada*) and a muscle attaches to the first adambulacral and the oral (*oradm*). The Y- and L-shaped circumoral ossicles are essentially modified ambulacrals (Mooi and David 2000), and the circumoral heads (*coh*) articulate across the mid-radial line by means of dentition. Transverse ambulacral muscle pairs (*abtam, actam*) are present on the circumorals, but are absent in *Calliasterella mira*.

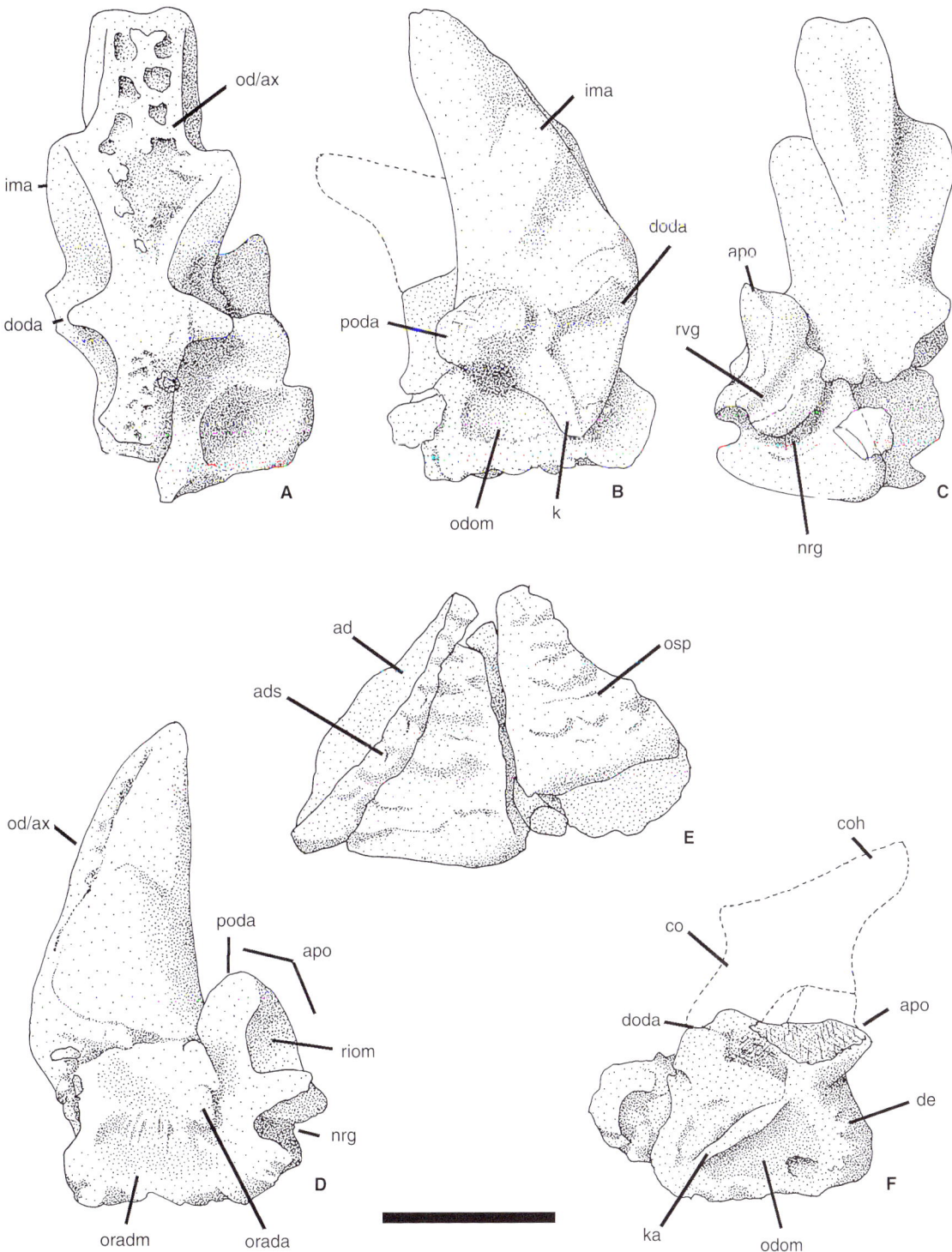

TEXT-FIG. 14. *Calliasterella mira*, mouth frame ossicles. Upper Carboniferous (Moscovian), Moscow, Russia. Specimen figured by Schöndorf (1909) (unregistered in PIN collections, Moscow). A–D, articulated axillary-oral ossicle in external (A), proximal (B), internal (C) and distal (D) aspects. E, pair of oral ossicles in actinal aspect. F, oral ossicle in internal (interradial) view. Scale bar 5 mm. See Table 2 for abbreviations.

The basic construction of the mouth frame is remarkably consistent in neoasteroids (Gale 1987) and differs most significantly in the highly derived Brisingida where the mouth frame forms an immobile, sometimes partially fused ring (e.g. G. O. Sars 1875; Fisher 1928; Pl. 12, figs 5–6 herein).

In abactinal, internal aspect, the mouth frame is pentagonal in outline in most species investigated, and the odontophores are deeply inset between the orals (e.g. *Asterina gibbosa*; Pl. 1, fig. 2; *Echinaster purpureus*; Pl. 11, fig. 2; *Ceramaster granularis* (Retzius, 1783) (Pl. 12, fig. 1). The frame is stellate with produced circumorals in the cribelline species *Ctenodiscus crispatus* and *Styracaster chuni* (Pl. 12, figs 2, 4). In both stellate and pentagonal constructions, the distal circumoral bar fits into a notch on the side of the odontophore (Pl. 1, fig. 2). In the forcipulate species *Asterias rubens*, *Heliaster helianthus*, *Freyella elegans* and *Brisinga costata*, the mouth frame is ring-like and the odontophores are shallower in position (Pl. 11, fig. 3; Pl. 12, figs 5–6). The mouth ring is inflexible in *Freyella elegans* (Pl. 12, figs 5–6) and *Brisinga costata*, and the odontophore and orals are fused in the latter.

Odontophore

The odontopore has barely been studied since the work of Viguier (1879), although it was described in detail in *Ctenodiscus crispatus* by Turner and Dearborn (1972). The odontophore (Text-fig. 13B, E) is made up of symmetrical, paired, distal and proximal processes (*poda*, *doda*), which articulate with the inner surfaces of the orals and an actinal keel (*k*) to which the fibres of the odontophore-oral muscle (*odom*) attach (Text-fig. 13C). In *Ctenodiscus crispatus*, the distal keel also articulates with the oral ossicle (Turner and Dearborn 1972). In *Calliasterella mira* (Text-fig. 14), the axillary/odontophore is vertically elongated, with a large, arrowhead-shaped external face that bears spines and articulates distally (*ima*) with the first marginals. The articulation structures on the actinal portion of the axillary in *Calliasterella mira* are homologous with those in neoasteroids and include equivalents of both proximal and distal odontophore articulations (*poda*, *doda*) and the keel (Text-fig. 15). A small oral-odontophore muscle was probably present and inserted into a cavity on the inner face of the oral. The odontophore is proportionately small in all neoasteroids, lacking

both an abactinal external process and any articulation with the marginals (Text-figs 13C, 15A). Certain paxillosid neoasteroids, including *Luidia ciliaris*, *Styracaster chuni*, *Benthopecten simplex* and *Ctenodiscus crispatus*, retain a small external spine-bearing face on the odontophore, but this is absent in all other taxa examined.

The odontophore has several distinct constructional forms in neoasteroids (Pl. 15). In *Luidia ciliaris* (Pl. 15, fig. 9), *Astropecten irregularis*, *Radiaster tizardi* and *Nardoa variolata* (Pl. 15, fig. 2), the plate is T-shaped, with the proximal bars forming a short transverse structure and the posterior part being narrower. In *Odontaster validus*, *Asterina gibbosa* (Pl. 15, fig. 11) and *Crossaster papposus*, the transverse bar is broad and the posterior articulation flattened. The plate is oval-rhomboidal and lacks prominent processes in *Pteraster pulvillus* (Pl. 15, fig. 5), *Remaster gourdoni* (Pl. 15, fig. 6), *Benthopecten simplex* (Pl. 15, fig. 3), *Echinaster purpureus* (Pl. 15, fig. 7), *Zoroaster fulgens* (Pl. 15, fig. 4), *Asterias rubens* (Pl. 15, fig. 10) and *Heliaster helianthus*. In these taxa, the odontophore sits like a cap (e.g. Pl. 15, fig. 5) on the orals and/or circumorals. In *Echinaster purpureus* (Pl. 15, fig. 7) and the forcipulatid taxa (*Asterias rubens*, *Zoroaster fulgens* and *Heliaster helianthus*), the odontophore articulates with both circumorals and orals, but in all other taxa it does so only with the orals. The odontophore is rectangular and convex externally in *Freyella elegans* and firmly articulated with the orals (Pl. 12, fig. 6; Pl. 14, fig. 3) and fused with the orals in *Brisinga costata*.

Oral ossicles

The oral ossicles are highly variable in form in neoasteroids. In the paxillosid taxa *Luidia ciliaris*, *Astropecten irregularis*, *Benthopecten simplex*, *Styracaster chuni*, *Ctenodiscus crispatus* and *Radiaster tizardi* (Text-fig. 16A–F), the apophyse is low and proportionately small, the proximal part short and the pairs of oral ossicles articulate by means of dentition. In taxa traditionally referred to valvatids (*Nardoa variolata*, *Mediaster aequalis*, *Asteropsis*

PLATE 11

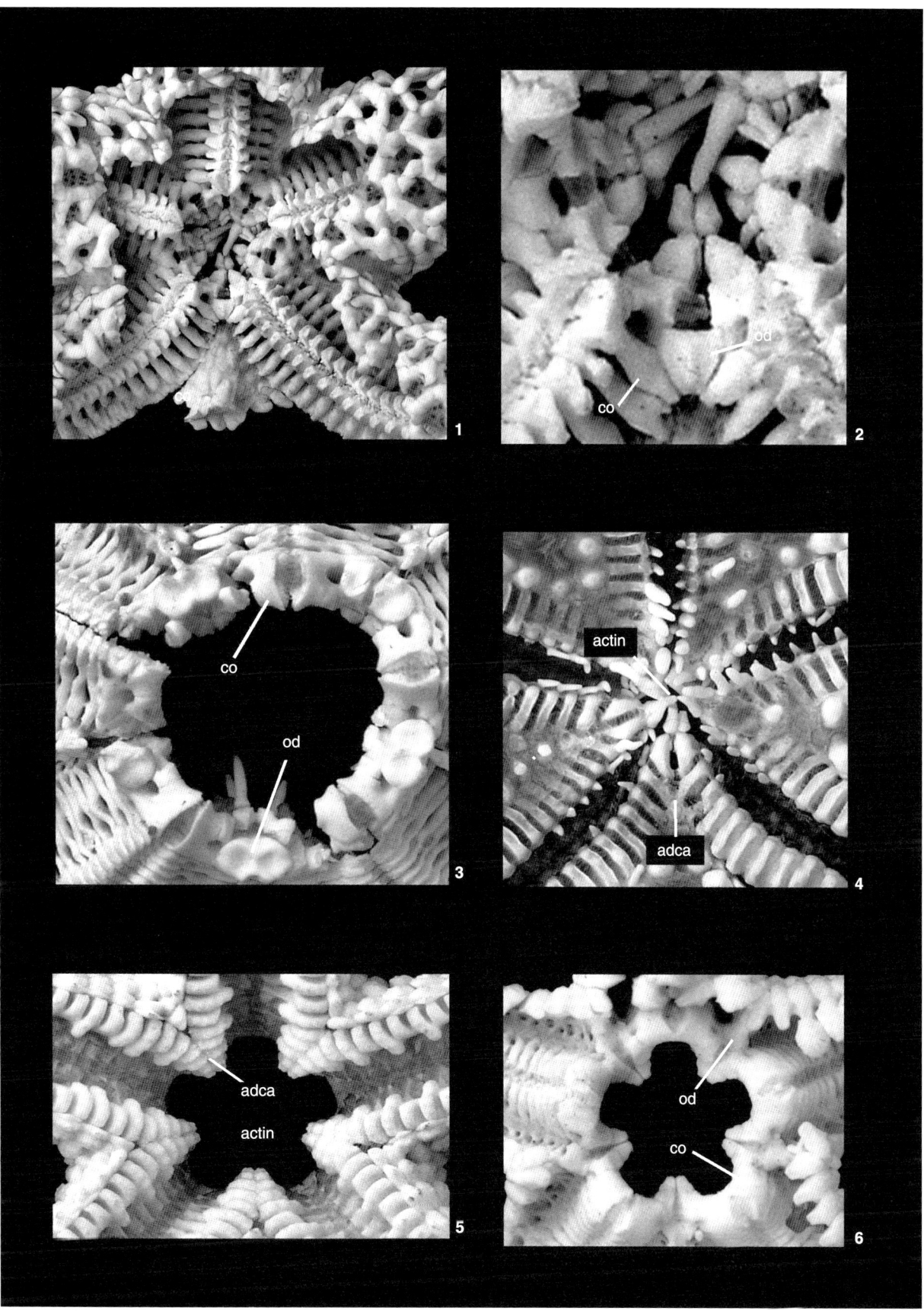

GALE, Mouth frame construction of Recent asteroids

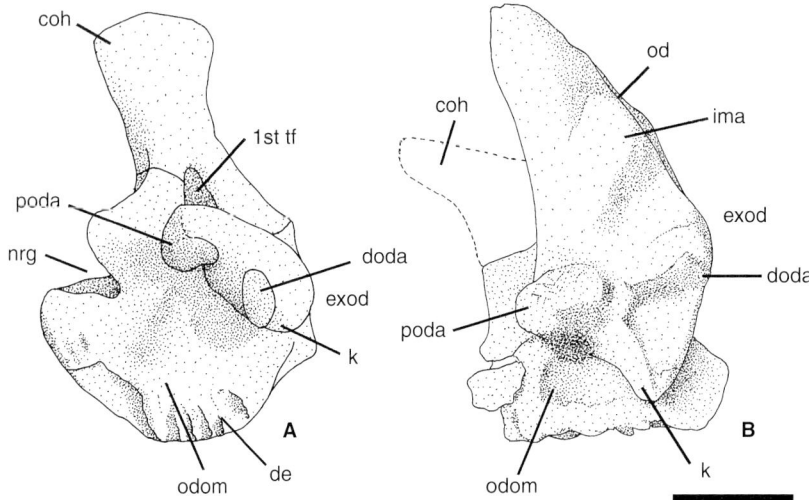

TEXT-FIG. 15. Mouth frame construction in *Luidia ciliaris* (A) and *Calliasterella mira* (B). Articulated odontophore/axillary, oral and circumoral ossicles in interradial aspect to show comparative morphology and homologies between Palaeozoic and post-Palaeozoic forms. Proximal left, distal right. Scale bar 5 mm. See Table 2 for abbreviations.

carinifera, Archaster lorioli, Acanthaster planci and *Protoreaster nodosus*; Text-fig. 16G–L), a prominent triangular proximal blade (*pb*) is present, and a number of matching interoral articulation structures composed of smooth peforate stereom occur on the inner (interradial) faces of the oral pairs (*iioa*; Pl. 14, fig. 8). The apophyse is large and inclined proximally in these taxa, and the oral-odontophore muscle inserts into a concavity on the distal part of the inner oral face, called an odontophore capsule (*odc*; Text-fig. 16G, J). The actinal faces are flattened, and the radial (outer) faces of the proximal blades of the orals articulate with each other, to make up a very closely articulated closure to the peristome (Pl. 12, figs 1, 3). A further group of taxa including *Asterina gibbosa, Pteraster pulvillus* and *Remaster gourdoni* (Pl. 13, figs 4–5, 9; Text-fig. 17A–F) share squarish orals with a short proximal blade and a very tall apophyse with a long and deep groove for the ring vessel. The outer (radial) faces of the orals of *Asterina gibbosa, Crossaster papposus, Remaster*

gourdoni, Pteraster pulvillus and *Odontaster validus* comprise a swollen proximal rim that carries the oral spines and a central, raised mound on which the suboral spines attach (Pl. 13, figs 4–5; Text-fig. 18B, C, E).

The oral ossicles of forcipulatid taxa (*Asterias rubens, Zoroaster fulgens, Heliaster helianthus, Freyella elegans* and *Brisinga costata*; Pl. 14, figs 1–6; Text-fig. 17G–J) are highly variable in shape, but all possess very short proximal blades and a tall, rounded proximal circumoral bar. The abactinal interradial interoral muscle pit (*abiim*) is large and deep, and the articulation surfaces for the adambulacrals are set at right angles to the main portion of the ossicles (Pl. 14, figs 1, 4, 6). The oral spine bases are few in number, and an actinostome (depressed, concave mouth region) is present. The first few adambulacrals of adjacent radii articulate together to form an adoral carina in many forcipulatid taxa (Pl. 11, fig. 5). It is interesting that a shallow actinostome and short adoral carina are present in *Echinaster purpureus* (Pl. 11, fig. 4).

EXPLANATION OF PLATE 12

Photographs of the dissected mouth frame of present-day asteroids in abactinal and actinal aspects.

Fig. 1. *Ceramaster granularis*, mouth frame in abactinal aspect. Note tight articulation of oral blades, closing peristome.

Fig. 2. *Ctenodiscus crispatus*, in abactinal aspect.

Fig. 3. *Nardoa variolata*, mouth frame in abactinal aspect.

Fig. 4. *Styracaster chuni*, mouth frame and disc in abactinal aspect. Note elongated circumoral ossicles.

Figs 5–6. *Freyella elegans*, mouth frame in actinal (5) and abactinal (6) views. Note ring-like form of mouth frame.

Abbreviations in Table 2.

Fig. 1, ×4; Fig. 2, ×4.5; Fig. 3, ×5.5; Fig. 4, ×5; Fig. 5, ×4; Fig. 6, ×3.5.

PLATE 12

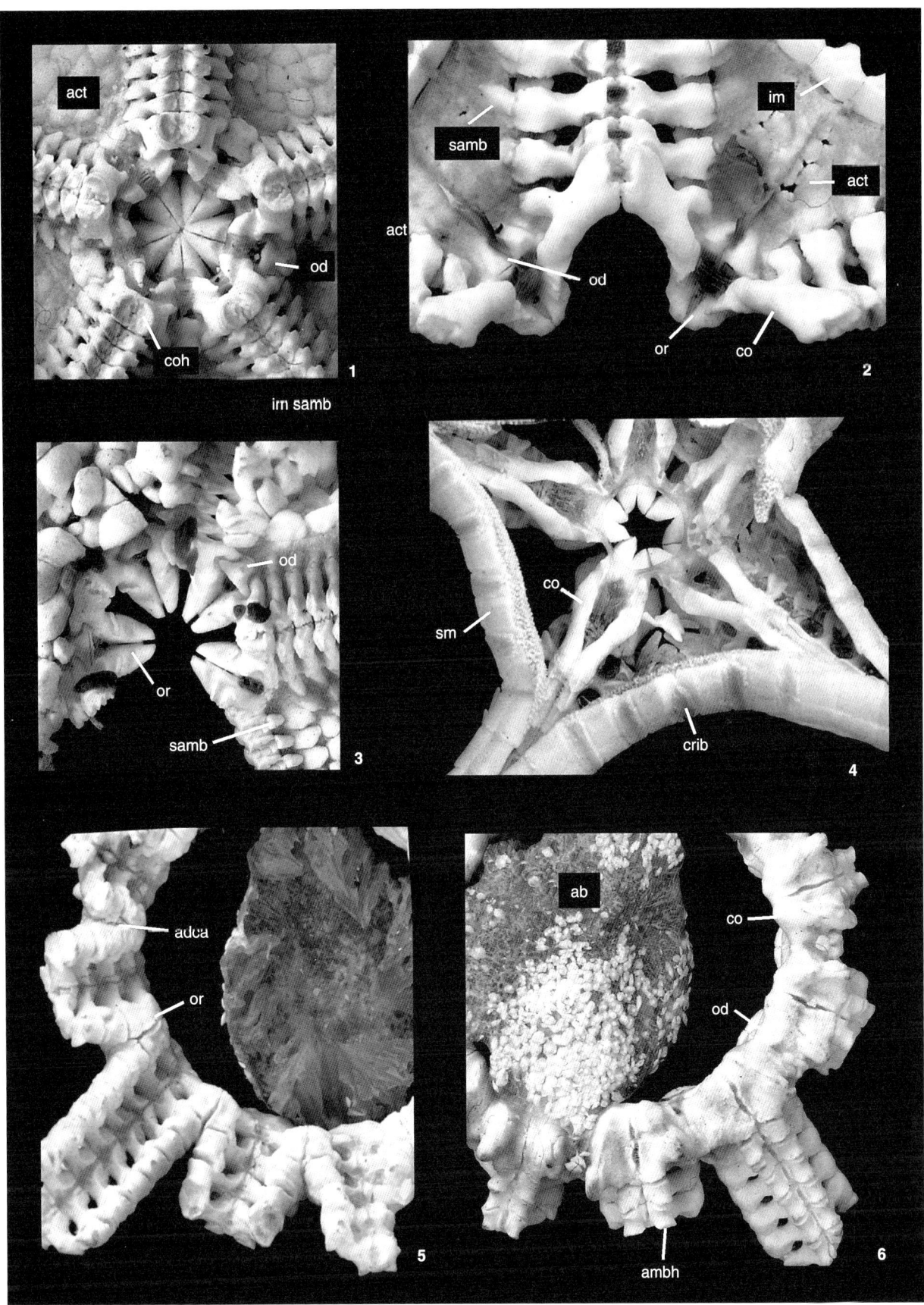

GALE, Mouth frame construction of Recent asteroids

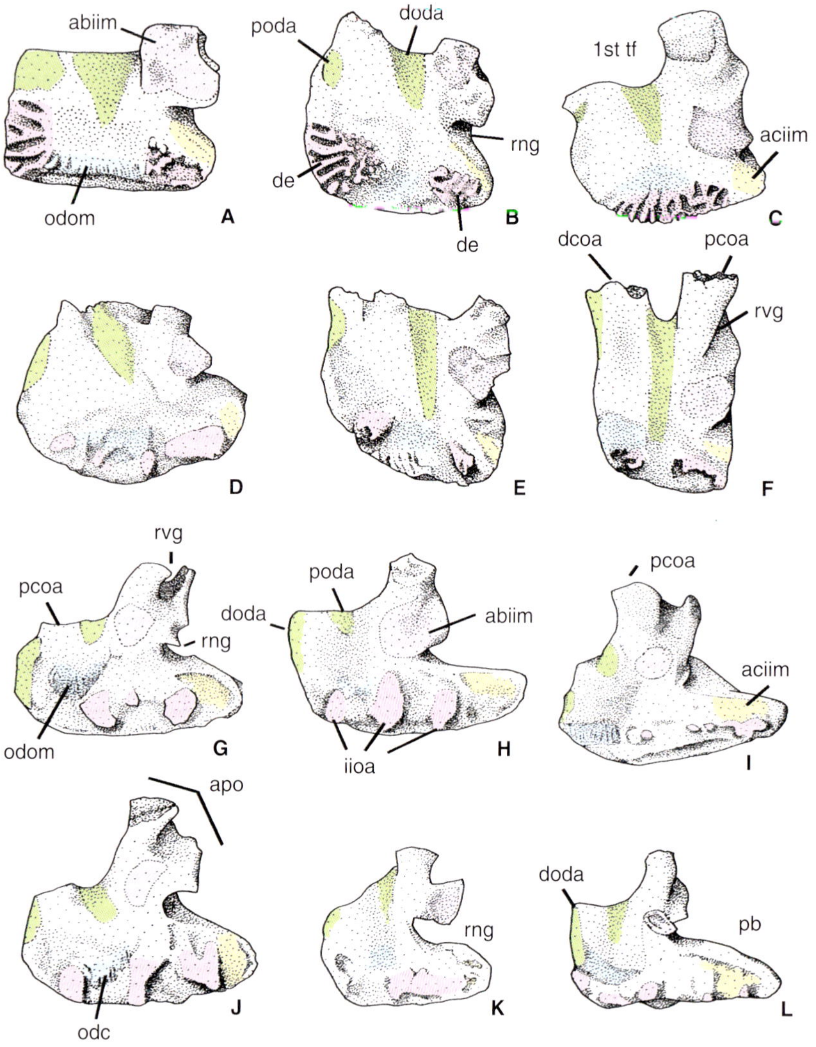

TEXT-FIG. 16. Drawings of dissociated oral ossicles in interradial aspect to show homologies of articulations, muscle insertion sites and structures that accommodate soft tissues. A, *Astropecten irregularis*. B, *Luidia ciliaris*. C, *Benthopecten simplex*. D, *Radiaster tizardi*. E, *Ctenodiscus crispatus*. F, *Styracaster chuni*. G, *Nardoa variolata*. H, *Asteropsis carinifera*. I, *Archaster lorioli*. J, *Mediaster aequalis*. K, *Acanthaster planci*. L, *Protoreaster nodosus*. Actinal to bottom of page, proximal to right. Not to scale. See Table 2 for abbreviations.

Circumorals

The circumoral ossicles are Y- or L-shaped in lateral aspect (Text-fig. 19), and the proximal circumoral process (*pcp*) articulating with the oral is short, the distal one (*dcp*) elongated. In paxillosids (*Luidia ciliaris*, *Astropecten irregularis*, *Radiaster tizardi*, *Styracaster chuni*, *Ctenodiscus crispatus* and *Benthopecten simplex*), the distal bars are strongly angled to the shaft of the ossicles (Text-fig. 19A–F); in all other taxa, the bar is more or less parallel with the shaft of the ossicle (Text-fig. 19G–V). A double articulation between the distal

---

EXPLANATION OF PLATE 13

SEM images of dissociated oral ossicles of neoasteroids; interradial (Figs 1, 3–4, 6–7, 9) and radial (Figs 2, 5, 8) views. Abactinal to top of page, actinal to bottom; proximal to right of page.

Figs 1–2. *Luidia ciliaris*.

Fig. 3. *Styracaster chuni*.

Figs 4–5. *Asterina gibbosa*.

Fig. 6. *Ctenodiscus crispatus*.

Figs 7–8. *Benthopecten simplex*.

Fig. 9. *Pteraster pulvillus*.

For abbreviations, see Table 2.

Scale bars represent 1 mm.

PLATE 13

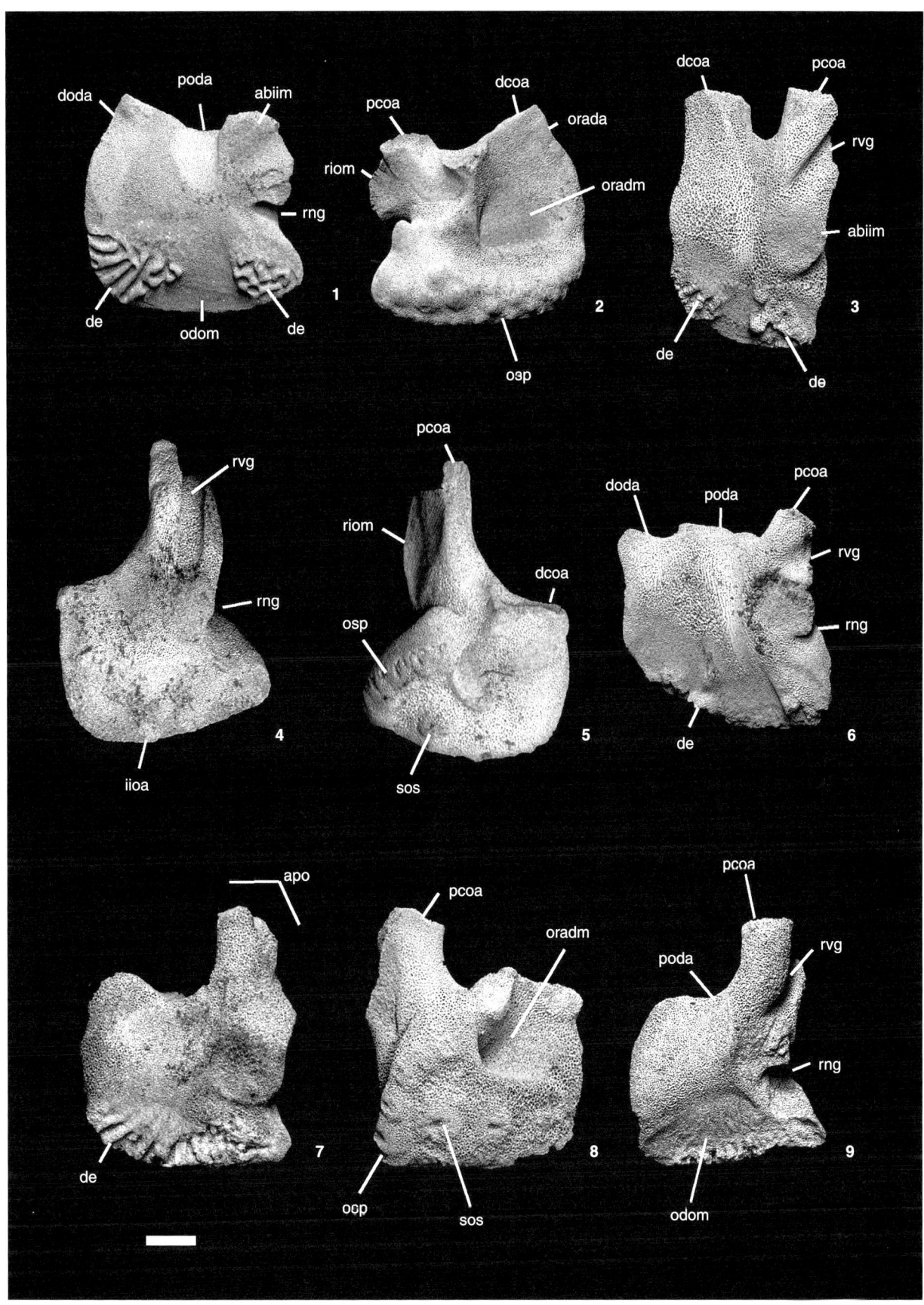

GALE, Morphology of oral ossicles of Recent asteroids

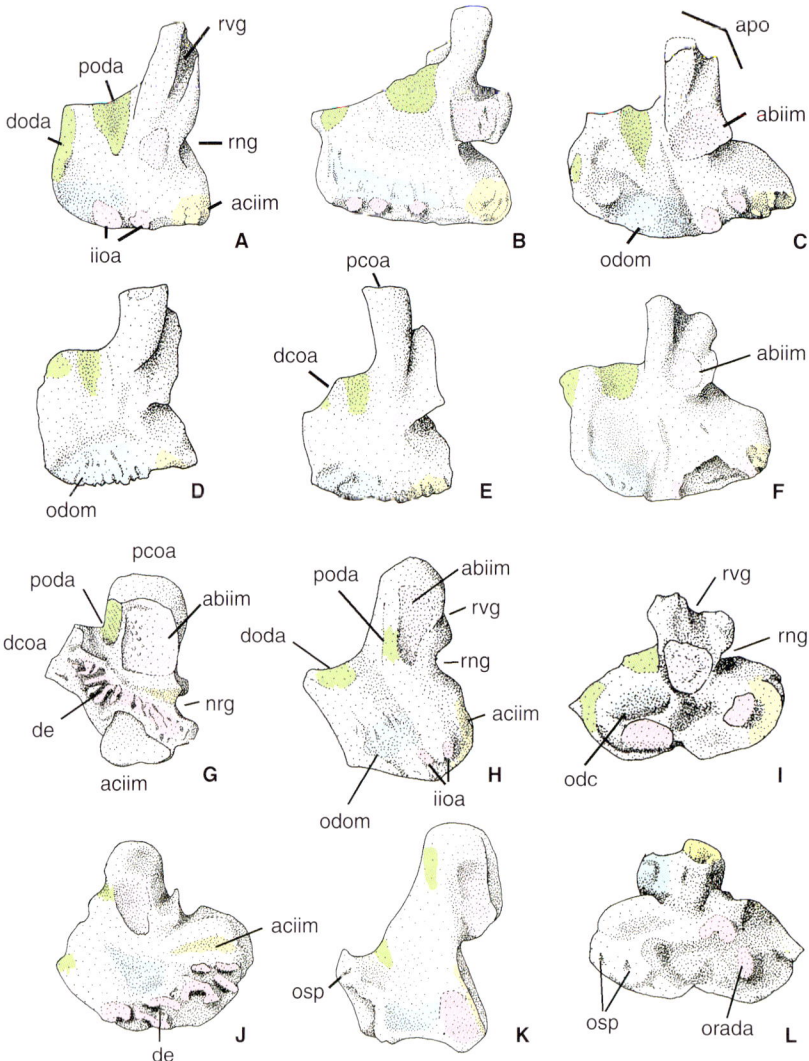

TEXT-FIG. 17. Drawings of dissociated oral ossicles in interradial aspect to show homologies of articulations, muscle insertion sites and structures, which accommodate soft tissues. A, *Asterina gibbosa*. B, *Crossaster papposus*. C, *Odontaster validus*. D, *Pteraster pulvillus*. E, *Remaster gourdoni*. F, *Porania pulvillus*. G, *Freyella elegans*. H, *Asterias rubens*. I, *Echinaster purpureus*. J, *Zoroaster fulgens*. K, *Heliaster helianthus*. L, *Echinaster purpureus* in radial view. Actinal to bottom of page, proximal to right of page. Not to scale. See Table 2 for abbreviations.

circumoral bar and the oral is clearly visible in the paxillosid taxa. The circumoral is as long as or greater than its total height in *Asterias rubens*, *Zoroaster fulgens*, *Freyella elegans* and *Brisinga costata* (Text-fig. 19A–D), but shorter than high in all other taxa. In *Freyella elegans* and *Brisinga costata*, the proximal articulation with

---

EXPLANATION OF PLATE 14

SEM images of dissociated oral ossicles of neoasteroids; interradial (Figs 2–3, 5, 8) and radial (Figs 1, 4, 6–7) views. Abactinal to top of page, actinal to bottom; proximal to right of page.

Figs 1–2. *Asterias rubens*.

Fig. 3. *Freyella elegans*.

Figs 4–5. *Zoroaster fulgens*.

Fig. 6. *Heliaster helianthus*.

Figs 7–8. *Nardoa variolata*.

See Table 2 for abbreviations.

Scale bar represents 1 mm.

PLATE 14

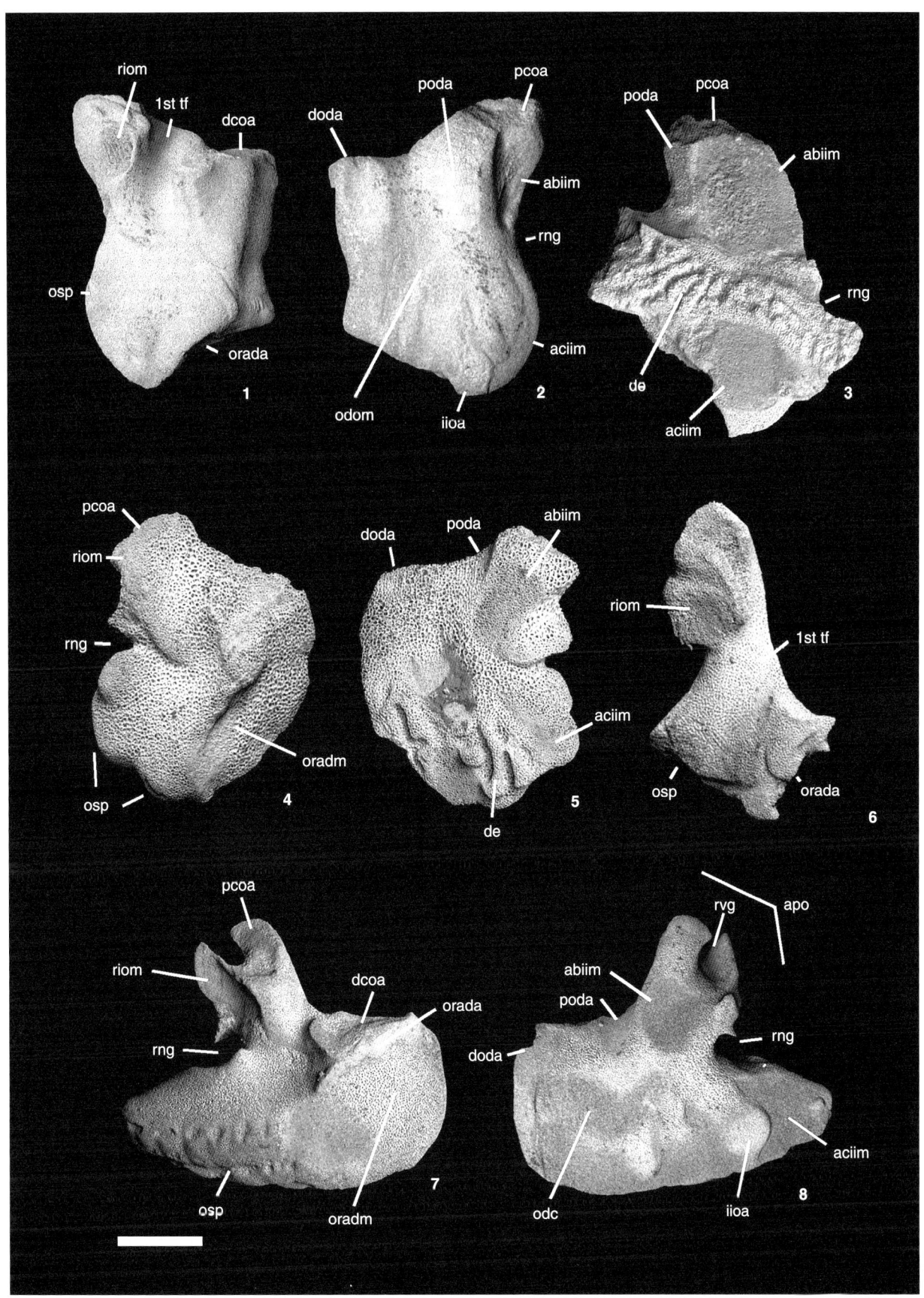

GALE, Morphology of oral ossicles of Recent asteroids

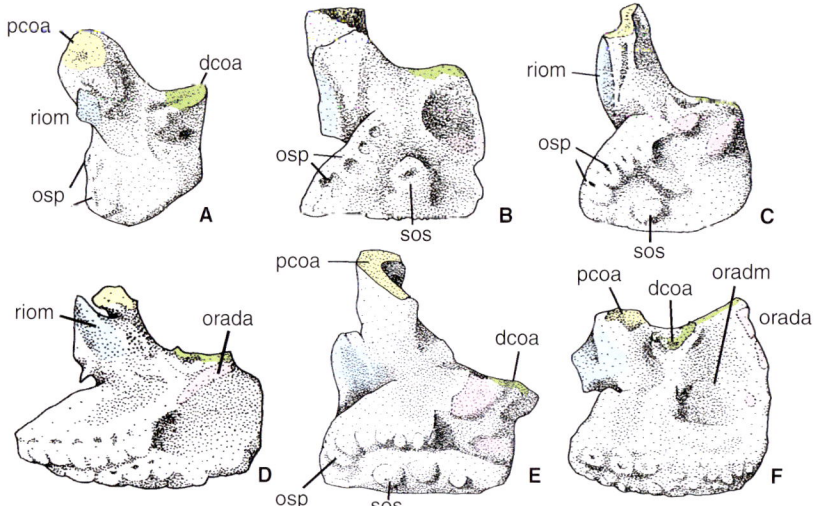

TEXT-FIG. 18. Drawings of dissociated oral ossicles in radial aspect to show homologies of articulations, muscle insertion sites and structures, which accommodate soft tissues. A, *Asterias rubens*. B, *Pteraster pulvillus*. C, *Asterina gibbosa*. D, *Nardoa variolata*. E, *Crossaster papposus*. F, *Luidia ciliaris*. Actinal to bottom of page, proximal to right of page. Not to scale. See Table 2 for abbreviations.

the oral is large and flat to concavo-convex (Text-fig. 19A, B).

COMPARATIVE MORPHOLOGY OF ASTEROID SOFT TISSUES

Digestive systems

The digestive system in asteroids has been described in detail by Jangoux (1982). Anus and rectum are absent in the paxillosids *Luidia ciliaris*, *Astropecten irregularis*, *Ctenodiscus crispatus* and *Styracaster chuni*, which also lack the ability to evert the stomach by retraction. All other neoasteroid species studied here (Table 1) have a rectum and anus and are capable of stomach eversion using a retractor system. Tiedemann's pouches, which are specializations of the oral part of the median duct of the pyloric caecae, are not found in paxillosids or forcipulatids, but are present in most other taxa studied here (Jangoux 1982). An elaborate pyloric structure called the pyloric complex is found in *Archaster lorioli*, *Echinaster purpureus*, *Odontaster validus*, *Mediaster aequalis*, *Protoreaster nodosus*, *Nardoa variolata* and *Asterina gibbosa*.

Tube feet

It was long supposed that asteroid tube feet could be divided simply into nonsuckered and suckered categories.

PLATE 15

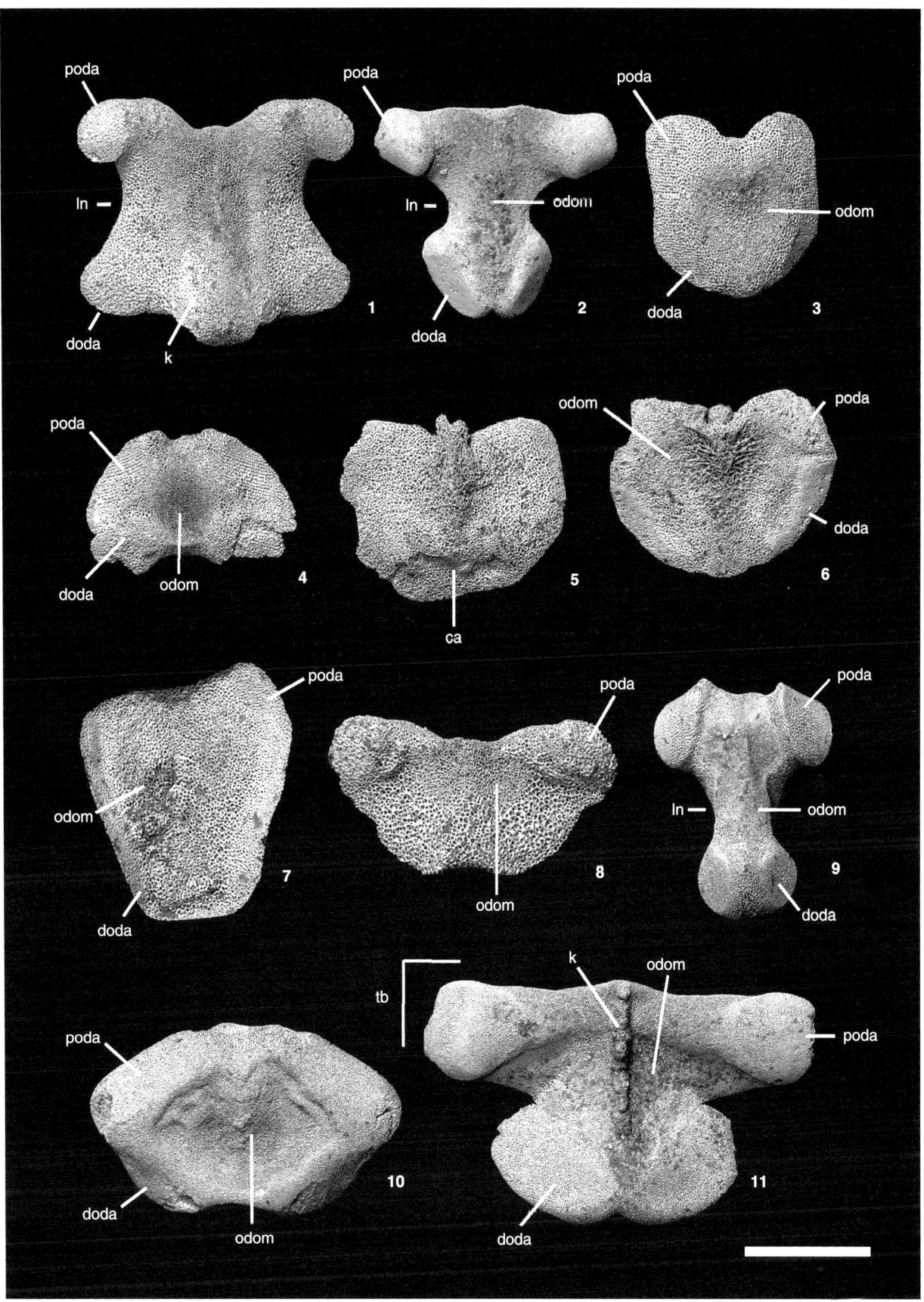

GALE, Morphology of odontophores of Recent asteroids

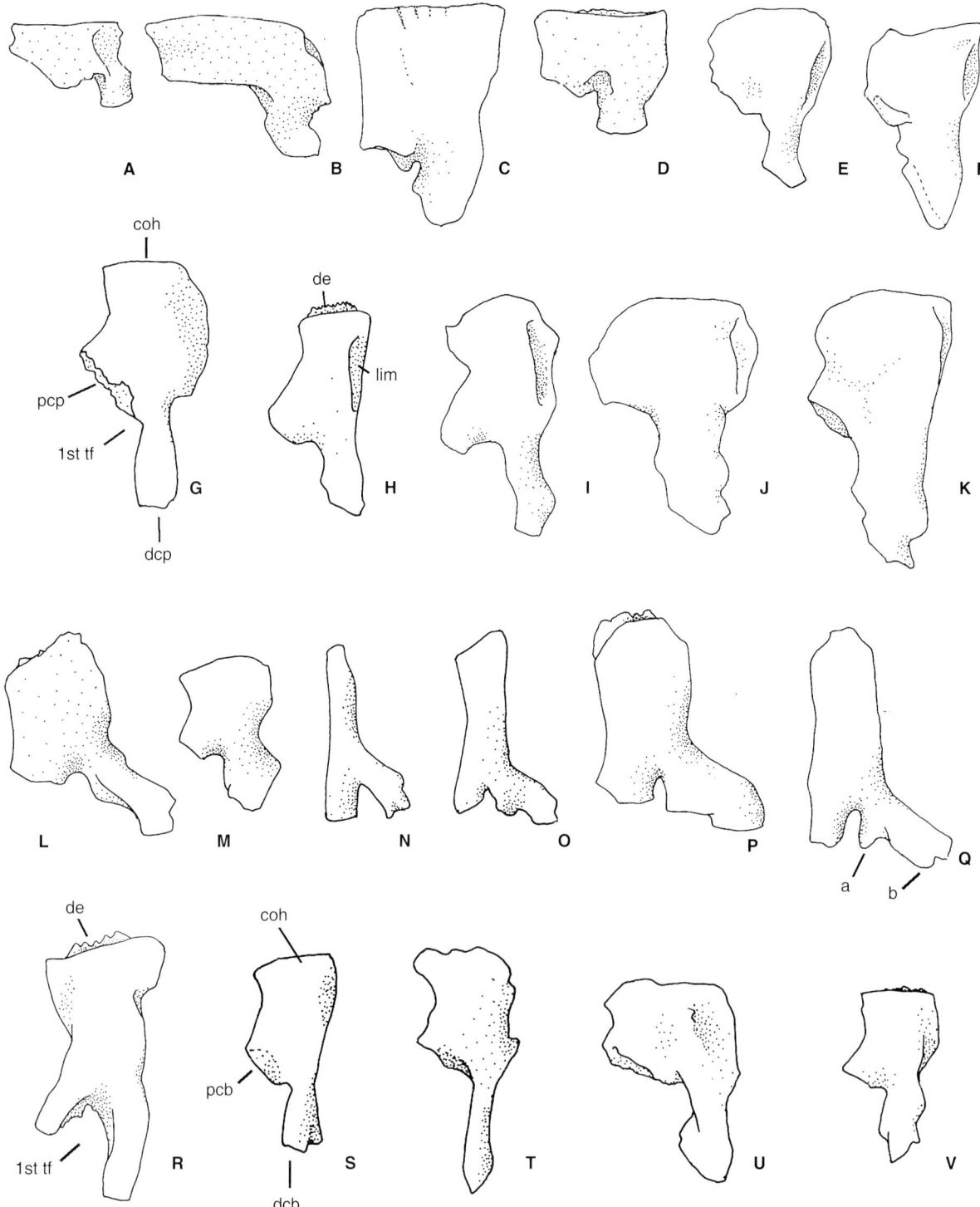

TEXT-FIG. 19. Dissociated circumoral ossicles of neoasteroids in lateral abradial view. A, *Freyella elegans*. B, *Brisinga costata*. C, *Zoroaster fulgens*. D, *Asterias rubens*. E, *Acanthaster planci*. F, *Echinaster purpureus*. G, *Pteraster pulvillus*. H, *Porania pulvillus*. I, *Protoreaster nodosus*. J, *Archaster lorioli*. K, *Nardoa variolata*. L, *Astropecten irregularis*. M, *Benthopecten simplex*. N, *Styracaster chuni*. O, *Ctenodiscus crispatus*. P, *Luidia ciliaris*. Q, *Radiaster tizardi*. R, *Odontaster validus*. S, *Asterina gibbosa*. T, *peribolaster biserialis*. U, *Crossaster papposus*. V, *Mediaster aequalis*. Abactinal to top of page, proximal to right of page. Not to scale. See Table 2 for abbreviations.

Nonsuckered tube feet are found exclusively in the paxillosids, and suckered ones occur in all other asteroids. Gale (1987) categorized the more derived asteroids with suckered tube feet as the Surculifera. More recently, Vickery and McClintock (2000) examined the histology of tube feet of 45 asteroids and derived a more complex classification, based on the shape of the terminal tube foot and the presence/absence of sucking discs.

Papulae

Papulae are small specialized respiratory structures that occur on the surface of asteroids, between ossicles. In paxillosids, *Asteropsis carinifera*, *Mediaster aequalis* and *Archaster lorioli* these are restricted to the interstices of the abactinal ossicles, but in other neoasteroid species these can occur between the marginals and actinal ossicles. In *Echinaster purpureus* and *Nardoa variolata*, the papular areas contain small, granule-like ossicles.

PHYLOGENETIC ANALYSIS

Methods

Previous cladistic studies of neoasteroid morphology (Gale 1987; Blake 1987; Blake *et al.* 2000; Blake and Hagdorn 2003) have suffered from a number of major limitations. The most significant of these is the presumption of polarity. Thus, for Gale (1987), the Paxillosida represented the primitive neoasteroid condition. In contrast, to Blake (1987, 1988*a*, *b*) the Paxillosida were secondarily simplified asteroids that had specialized to living on soft substrates. Current cladistic procedure uses simultaneous unconstrained analysis in which polarity is obtained only with reference to an outgroup (Kitching *et al.* 1999). Blake effectively presumed the monophyly of major taxonomic groups and analysed these without reference to any outgroups (Text-fig. 20). This has significant implications if the groups are actually paraphyletic. For example, Blake's (1987) analysis placed the goniasterid subfamily Pseudarchasterinae Sladen, 1889 as sister group to all other Valvatida, which suggests that the Goniasteridae Forbes, 1841 at least are paraphyletic.

Second, data sets used by both Gale (1907) and Blake (1987, 1988*b*) included gradational and proportional characters that are impossible to code accurately. Gale did not use PAUP but rather constructed the cladogram by hand. Third, both authors used families as terminal taxa, problematic because certain of these are probably paraphyletic and in any case can show a considerable range of morphology between constituent genera.

The character set for neoasteroids (Appendix) thus aims to provide comprehensive coverage of the morphology (gross and ossicle morphology, tube feet histology, digestive system morphology) of 24 extant species that have traditionally been classified in as many families. *Calliasterella mira* from the Upper Carboniferous of Moscow is the default outgroup. This species is redescribed below. Character states are shown in Table 6.

Results

Heuristic, unconstrained simultaneous analysis of the dataset using PAUP 4, optimized to deltran,

TEXT-FIG. 20. Strict consensus cladogram from six best trees, unconstrained simultaneous analysis, default outgroup, optimized to deltran. Changes in character states (see Appendix) shown.

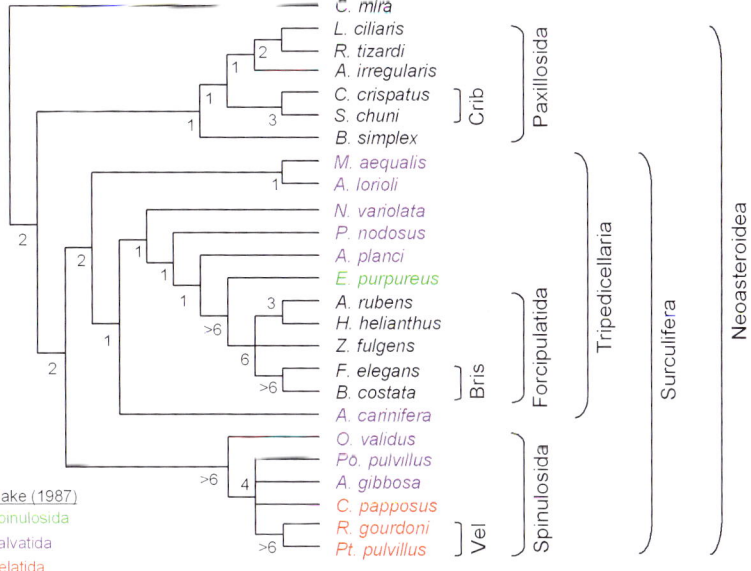

Blake (1987)
Spinulosida
Valvatida
Velatida

TABLE 6. Character matrix used in cladistic analysis.

	1	2	3	4	5	6	7	8	9	10	11	12	13	14	15	16	17	18	19	20	21	22	23	24
C. mira	0	2	0	0	0	0	0	0	0	?	?	?	?	?	?	0	0	?	0	0	0	0	0	0
L. ciliaris	1	1	0	2	0	0	0	0	0	0	0	0	0	0	0	0	1	0	0	0	0	1	0	0
A. irregularis	0	1	0	2	0	0	0	0	0	0	0	0	0	0	0	0	1	0	0	0	0	1	0	0
R. tizardi	0	1	1	0	0	0	0	0	0	0	0	0	0	0	0	0	1	0	0	0	0	1	0	0
C. crispatus	0	1	1	2	0	0	0	0	0	0	0	0	0	0	0	0	1	0	0	0	0	1	0	0
S. chuni	0	1	1	2	0	0	0	0	0	0	0	0	0	0	0	2	0	0	0	0	0	1	0	0
B. simplex	0	1	0	2	0	0	0	0	0	0	0	1	0	0	0	2	0	0	0	0	0	1	0	0
M. aequalis	0	1	1	2	0	1	0	0	0	1	0	1	1	1	1	1	1	0	0	0	0	1	0	0
A. lorioli	0	1	0	2	0	0	2	0	0	1	0	1	1	1	1	1	1	0	0	0	0	1	0	0
O. validus	0	1	1	0	1	0	0	0	0	1	0	1	1	1	0	1	1	0	0	0	0	1	0	0
N. variolarius	0	2	0	0	0	1	0	0	0	1	0	1	1	1	1	1	1	0	1	1	0	0	0	0
P. nodosus	0	1	0	1	0	1	2	0	0	1	0	1	1	1	1	0	1	0	1	1	0	0	0	0
P. pulvillus	0	0	1	1	1	0	1	0	0	1	0	1	1	1	?	1	1	0	0	1	0	0	0	0
A. carinifera	0	1	1	1	0	0	0	0	0	1	0	1	1	1	1	1	1	0	0	0	0	0	0	0
A. gibbosa	0	0	1	1	1	0	1	0	0	1	0	1	1	1	0	1	1	0	1	1	0	0	0	0
A. planci	2	1	0	0	0	0	2	0	0	1	0	1	1	1	1	1	1	0	1	1	1	0	1	1
E. purpureus	0	2	0	0	0	0	0	0	0	1	1	1	1	1	1	1	1	1	1	1	1	0	1	1
C. papposus	2	1	0	1	1	0	1	0	0	1	1	1	1	1	0	1	1	0	1	1	0	1	0	0
R. gourdoni	0	0	1	1	1	0	–	1	1	1	1	1	1	1	0	1	1	0	1	1	0	1	0	0
Pt. pulvillus	0	0	1	1	1	0	–	1	1	1	1	1	1	1	0	1	1	0	1	1	0	1	0	0
A. rubers	0	2	0	0	0	1	2	0	0	1	1	1	1	0	0	1	1	1	1	1	1	0	1	1
Z. fulgens	0	2	0	0	0	0	2	0	0	1	1	1	1	0	0	0	1	0	0	1	1	0	1	1
H. helianthus	2	1	0	0	0	1	2	0	0	1	1	1	1	0	0	1	1	1	1	1	1	0	1	1
F. elegans	1	2	0	0	1	–	–	0	0	1	1	1	1	0	0	2	1	1	0	0	0	0	0	0
B. costata	2	2	0	0	1	–	–	0	0	1	1	1	1	0	0	2	0	1	0	0	0	0	0	0

	25	26	27	28	29	30	31	32	33	34	35	36	37	38	39	40	41	42	43	44	45	46	47	48
C. mira	?	?	?	0	0	0	1	1	1	0	0	0	0	–	–	0	0	0	–	–	–	–	–	–
L. ciliaris	0	0	0	0	0	0	0	1	1	0	0	1	1	1	–	1	1	0	–	0	0	0	0	1
A. irregularis	0	0	0	1	0	1	0	1	1	0	0	1	1	1	0	1	0	0	–	–	–	–	–	–
R. tizardi	0	0	0	0	0	1	0	0	1	1	0	1	1	1	0	1	0	0	–	–	–	–	–	–
C. crispatus	0	0	0	1	0	1	0	1	1	0	0	2	1	1	0	1	0	0	–	–	–	–	–	–
S. chuni	0	0	0	0	0	1	0	1	1	0	0	2	1	0	0	0	0	0	–	–	–	–	–	–
B. simplex	0	0	0	0	0	0	1	0	0	0	1	1	0	0	0	1	1	0	?	?	?	0	?	?
M. aequalis	0	0	0	0	0	1	0	1	1	0	0	0	1	0	0	0	1	0	1	0	1	1	0	1
A. lorioli	1	0	0	0	0	1	0	1	1	0	0	0	1	0	0	0	1	0	1	0	1	1	0	1
O. validus	0	0	0	0	0	1	0	0	0	1	0	0	1	1	0	0	0	0	–	–	–	–	–	–
N. variolarius	2	1	1	0	0	1	1	0	0	0	1	0	1	0	0	0	0	0	–	–	–	–	–	–
P. nodosus	2	1	1	0	0	1	1	1	1	0	1	0	1	0	0	0	1	0	1	1	1	1	0	1
P. pulvillus	0	1	0	0	0	1	1	0	0	0	0	0	1	1	1	0	0	0	–	–	–	–	–	–
A. carinifera	2	1	1	0	0	1	1	1	0	0	0	0	1	0	0	0	0	0	–	–	–	–	–	–
A. gibbosa	0	1	0	0	0	1	1	0	0	0	0	0	1	1	0	0	0	0	–	–	–	–	–	–
A. planci	2	1	1	0	0	1	1	0	0	0	1	0	1	0	–	0	1	0	1	1	1	1	0	1
E. purpureus	2	1	1	0	0	1	1	0	0	0	1	0	1	0	0	0	0	0	–	–	–	–	–	–
C. papposus	0	1	0	0	0	1	1	0	0	1	0	0	1	1	–	0	0	0	–	–	–	–	–	–
R. gourdoni	0	1	0	0	1	1	1	0	0	0	0	0	1	–	–	0	0	0	–	–	–	–	–	–
Pt. pulvillus	0	1	0	0	1	1	1	0	0	0	0	0	1	–	–	0	0	0	–	–	–	–	–	–
A. rubers	2	1	0	0	0	1	1	0	0	0	1	0	1	0	0	0	1	0	1	1	1	1	1	0
Z. fulgens	2	1	0	0	0	0	1	0	0	0	1	0	1	0	0	0	1	0	1	1	1	1	1	0
H. helianthus	2	1	0	0	0	1	1	0	0	0	?	0	1	0	–	0	1	0	1	1	1	1	1	0
F. elegans	–	–	0	0	0	1	1	0	0	0	?	0	1	0	–	0	1	0	1	1	1	1	1	0
B. costata	–	–	0	0	0	1	?	0	0	0	?	0	1	0	–	0	1	0	1	1	1	1	1	0

TABLE 6. Continued

	49	50	51	52	53	54	55	56	57	58	59	60	61	62	63	64	65	66	67	68	69	70	71	72
C. mira	–	–	–	–	–	0	0	0	0	0	0	0	0	0	0	0	0	1	1	0	0	0	0	0
L. ciliaris	1	–	–	–	–	1	1	0	0	1	0	0	0	0	0	0	0	1	1	1	1	0	0	0
A. irregularis	–	–	–	–	–	1	1	0	0	1	0	0	0	0	0	0	0	1	1	0	1	0	0	0
R. tizardi	–	–	–	–	–	1	1	0	0	1	0	0	0	0	0	0	0	1	1	1	1	0	0	0
C. crispatus	–	–	–	–	–	1	1	0	0	1	0	0	0	0	0	0	0	1	1	0	1	0	0	1
S. chuni	–	–	–	–	–	1	1	0	0	1	0	0	0	0	0	0	0	1	1	0	0	0	0	1
B. simplex	0	–	–	–	–	1	1	0	0	1	0	0	0	0	0	0	0	1	1	0	0	0	0	0
M. aequalis	1	–		–		1	1	0	0	1	1	0	0	0	0	0	0	1	0	0	1	0	1	0
A. lorioli	1	–	–	–	–	1	1	0	0	1	1	0	0	0	0	0	0	1	0	0	0	0	1	0
O. validus	–	–	–	–	–	1	1	0	0	1	2	0	0	1	0	0	1	0	0	0	0	0	0	0
N. variolarius	–	–	–		–	1	1	0	0	1	1	0	0	0	0	0	0	0	0	0	1	0	1	0
P. nodosus	1	–	0	–	–	1	1	0	0	1	1	0	0	0	0	0	0	0	0	0	0	0	1	0
P. pulvillus	–	–	–	–	–	1	1	0	0	1	2	1	1	2	0	1	1	0	0	0	0	0	1	0
A. carinifera	–	–	–	–	–	1	1	0	0	1	1	0	0	2	0	0	0	0	0	0	0	0	1	0
A. gibbosa	–	–	–	–	–	1	1	0	0	1	2	1	1	2	0	1	1	0	0	0	0	0	1	0
A. planci	1	–	0	–	–	1	1	0	0	1	2	1	1	2	0	0	0	0	0	0	0	0	1	0
E. purpureus	–	–	–	–	–	1	1	0	0	1	1	0	0	1	0	1	0	0	0	0	0	0	1	0
C. papposus	–	–	–	–	–	1	1	0	0	1	2	1	1	2	0	1	1	0	0	0	0	0	1	0
R. gourdoni	–	–	–	–	–	1	1	0	0	1	2	2	1	2	0	1	1	0	0	0	0	0	1	0
Pt. pulvillus	–	–	–	–	–	1	1	0	0	1	2	2	1	2	0	1	1	0	0	0	0	0	1	0
A. rubers	0	1	1	1	1	1	1	0	0	1	1	0	0	2	0	0	1	0	0	0	0	0	1	1
Z. fulgens	0	0	0	0	–	1	1	0	0	1	1	0	0	1	0	0	0	0	0	0	0	0	1	0
H. helianthus	0	1	1	2	1	1	1	0	0	1	1	0	0	1	0	0	1	0	0	0	0	0	1	1
F. elegans	0	1	1	2	1	1	1	1	1	1	3	0	0	0	1	0	0	0	0	0	0	0	1	0
B. costata	0	1	1	2	1	1	1	1	1	1	3	0	0	0	1	0	0	0	0	0	0	0	1	0

	73	74	75	76	77	78	79	80	81	82	83	84	85	86	87	88	89	90	91	92	93	94	95	96
C. mira	0	0	0	0	0	0	0	0	0	0	0	0	0	0	0	0	0	1	0	0	4	0	0	0
L. ciliaris	0	0	1	1	0	0	0	0	0	0	0	0	0	0	0	0	0	1	0	0	0	0	1	1
A. irregularis	0	0	1	1	0	1	0	0	0	0	0	1	0	0	0	0	0	1	0	0	0	0	1	1
R. tizardi	0	0	1	1	0	0	0	0	0	0	0	1	0	0	0	0	0	1	0	0	0	0	1	1
C. crispatus	1	1	1	1	0	1	0	0	0	0	0	1	0	1	0	0	0	1	0	0	0	0	1	1
S. chuni	1	1	1	1	0	1	0	0	0	0	0	1	0	1	0	0	0	1	0	0	0	0	1	1
B. simplex	0	1	1	1	0	1	0	0	0	0	0	1	0	0	0	0	0	1	0	0	2	0	1	1
M. aequalis	0	1	1	1	0	2	0	0	0	0	0	1	0	0	0	0	0	0	0	0	0	0	1	1
A. lorioli	0	1	1	1	0	2	0	0	0	0	0	1	0	0	0	0	0	0	0	0	0	0	1	1
O. validus	0	1	1	1	1	2	1	0	0	0	0	1	0	0	0	0	0	1	0	0	1	0	1	1
N. variolarius	0	1	1	1	0	2	0	0	0	0	0	1	0	0	0	0	0	0	0	0	0	0	1	1
P. nodosus	0	1	1	1	0	2	0	0	0	0	0	1	0	0	0	0	0	0	0	0	0	0	1	1
P. pulvillus	0	1	1	1	1	2	1	0	0	0	0	1	0	0	0	0	0	1	0	0	1	0	1	1
A. carinifera	0	1	1	1	0	2	0	0	0	0	0	1	0	0	0	0	0	0	0	0	0	0	1	1
A. gibbosa	0	1	1	1	0	2	0	0	0	0	0	1	0	0	0	0	0	1	0	0	1	0	1	1
A. planci	0	1	1	1	0	2	0	0	0	0	0	1	0	0	0	0	0	1	0	0	0	0	1	1
E. purpureus	0	1	1	1	0	2	0	0	0	0	0	0	1	0	0	1	1	2	1	1	2	1	1	1
C. papposus	0	1	1	1	1	2	1	0	0	0	0	1	0	0	0	0	0	1	0	0	1	0	1	1
R. gourdoni	0	1	1	1	1	2	1	1	1	1	0	1	0	0	0	0	0	1	0	0	2	1	1	1
Pt. pulvillus	0	1	1	1	1	2	1	1	1	1	0	1	0	0	0	0	0	1	0	0	2	1	1	1
A. rubers	0	1	1	1	0	2	0	0	0	0	0	0	1	2	1	1	1	2	1	1	2	1	1	1
Z. fulgens	0	1	1	1	0	2	0	0	0	0	0	0	1	0	1	1	1	2	1	1	2	1	1	1
H. helianthus	0	1	1	1	0	2	0	0	0	0	0	0	1	2	1	1	1	2	1	1	2	1	1	1
F. elegans	0	1	1	1	0	2	0	0	0	0	1	0	1	3	2	1	1	2	1	1	3	2	1	1
B. costata	0	1	1	1	0	2	0	?	0	0	1	0	1	3	2	1	1	2	1	1	3	2	1	1

TABLE 6. Continued

	97	98	99	100	101	102	103	104	105	106	107	108	109	110	111	112	113	114	115	116	117	118	119	120
C. mira	0	0	0	0	0	0	0	0	1	1	0	0	0	0	1	0	?	0	0	?	0	0	0	0
L. ciliaris	0	0	0	0	0	0	0	0	0	0	0	0	0	0	0	0	1	0	0	0	0	0	0	0
A. irregularis	1	0	0	0	0	0	0	0	0	0	0	0	0	0	0	0	1	0	0	0	0	0	0	0
R. tizardi	0	0	0	0	0	0	0	0	0	0	0	0	0	0	0	0	1	0	0	0	0	0	0	0
C. crispatus	0	0	0	0	0	0	0	0	0	0	0	1	0	0	0	0	1	0	0	0	0	0	0	0
S. chuni	0	0	0	0	0	0	0	0	0	0	0	1	0	0	0	0	1	0	0	0	0	0	0	0
B. simplex	0	0	0	0	1	0	0	0	0	0	0	0	0	0	0	0	1	0	0	0	0	0	0	0
M. aequalis	1	0	0	0	0	0	0	1	1	1	0	0	0	0	0	0	2	1	1	0	0	1	0	0
A. lorioli	1	0	0	0	0	0	0	1	1	1	0	0	0	0	0	0	2	1	1	0	0	1	0	0
O. validus	1	0	0	0	0	0	0	0	1	1	0	0	0	0	0	0	0	1	0	0	1	0	0	0
N. variolarius	1	0	0	0	0	0	0	1	1	1	0	0	0	0	0	0	2	1	1	0	0	1	0	0
P. nodosus	1	0	0	0	0	0	0	1	1	1	0	0	0	0	0	0	2	1	1	0	0	1	0	0
P. pulvillus	1	0	0	0	0	0	0	0	1	1	0	0	0	0	0	0	0	1	0	0	1	0	0	0
A. carinifera	1	0	0	0	0	0	0	1	1	1	0	0	0	0	0	0	2	1	1	0	0	1	0	0
A. gibbosa	1	0	0	0	0	0	0	0	1	1	0	0	0	0	0	0	0	1	0	0	1	0	0	0
A. planci	1	0	0	0	0	0	0	1	1	1	0	0	0	0	0	0	2	1	1	0	0	1	0	0
E. purpureus	1	0	1	0	1	1	0	1	1	1	0	0	0	0	0	0	2	1	0	0	0	0	0	0
C. papposus	1	0	0	0	0	0	0	0	1	1	0	0	0	0	0	0	0	1	0	0	0	0	0	0
R. gourdoni	1	0	0	0	0	0	1	0	1	1	0	0	0	0	0	0	0	1	0	0	1	0	0	0
Pt. pulvillus	1	0	0	0	1	1	1	0	1	1	0	0	0	0	0	0	0	1	1	0	1	0	0	0
A. rubers	1	0	1	0	1	1	0	1	1	1	1	0	0	0	1	0	0	1	1	1	0	0	1	1
Z. fulgens	1	0	1	0	1	1	0	1	1	1	1	0	0	0	1	0	1	1	1	1	0	0	1	1
H. helianthus	1	0	1	0	1	1	0	1	1	1	1	0	0	0	1	0	0	1	1	1	0	0	1	1
F. elegans	1	1	1	1	1	1	0	0	1	1	1	1	0	1	1	1	1	1	1	1	0	0	1	1
B. costata	1	2	1	1	1	1	0	0	1	1	1	1	0	1	1	1	1	1	1	1	0	0	1	1

generated six best trees, and the consensus tree of these is illustrated (Text-fig. 21). The variation in the six trees is small and involves rearrangement of taxa within the Spinulosida and Forcipulatida, which appear as polytomies in the consensus tree. In the following section, synapomorphies are indicated by an asterisk (*). It is important to note that defining character states refer only to the taxa under analysis in the present study.

Phylogenetic conclusions and taxonomic recommendations

Class ASTEROIDEA de Blainville, 1830

Following Shackleton (2005), the asteroids can be defined by the presence of directly opposing ambulacral ossicles, blocky adambulacrals that are positioned actinally to the ambulacrals, longitudinally adjacent ambulacrals which abut, the presence of inferomarginal spines, a madreporite abactinal in position to the inferomarginals, the aboral plating in regular longitudinal rows and the stellate body shape.

Clade NEOASTEROIDEA Gale, 1987

 30 (0–1) *superomarginals present (absent in Zoroaster fulgens, Luidia ciliaris)*
*37 (0–1) *actinal ossicles present*
*54 (0–1) *abactinal transverse ambulacral muscles present*
*55 (0–1) *actinal transverse ambulacral muscle present*
*58 (0–1) *ambulacral base <50 per cent of total breadth of ossicle*
*75 (0–1) *adambulacrals imbricate proximally*
*76 (1–0) *adambulacrals equally broad as or narrower than ambulacrals*
 84 (0–1) *discrete adambulacral and subadambulacral spines present*
*93 (4–0) *odontophore/axillary lacks elongated external face*
*95 (0–1) *odontophore/axillary does not articulate with marginals*
*96 (0–1) *odontophore/axillary low, flattened*

The Neoasteroidea are a monophyletic, well-characterized Triassic to Recent group and constitute the crown group of the Asteroidea, with Palaeozoic taxa forming the stem group. The more crownward part of the stem group occurs in Upper Palaeozoic rocks, but their skeletal morphology is poorly known in taxa other than the Late Carboniferous *Calliasterella mira*, the outgroup taxon

TEXT-FIG. 21. Plot of the relative position of the madreporite to the centre of the disc and inner margin of the interradial superomarginal (PM ratio) based on 112 living and fossil species. The boxes represent the lower to upper quartile ranges, the vertical bars the total ranges of values. Mean values are significantly different (Student's *t*-test). There is a clear division between paxillosids in which the madreporite is more marginal (PM > 1) and surculiferans in which it is more central (PM < 1), which correlates with different modes of early postmetamorphosis development of the skeleton (Text-fig. 32). Palaeozoic species have even more marginal madreporites than neoasteroids.

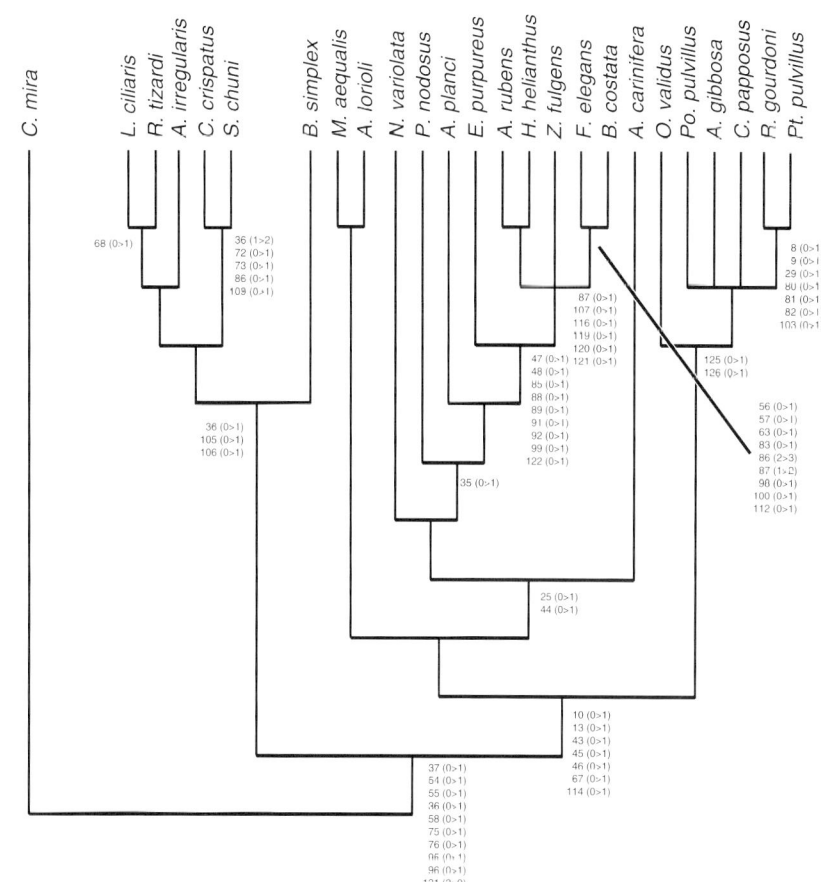

used in the present study. *Permaster grandis* Kesling, 1969 from the Permian of Western Australia is moderately well preserved, possesses both transverse ambulacral muscles and lacks a tall external process on the odontophore/axillary. This species is probably closer to the neoasteroids than is *Calliasterella mira*, but awaits redescription. High-resolution understanding of the relationships of neoasteroids and Late Palaeozoic taxa will be assisted by the discovery of well-preserved ambulacral groove and mouth frame ossicles.

Clade PAXILLOSIDA Perrier, 1884

 4 (0–2) *arm rectangular in cross-section (paralleled in Archaster lorioli, Mediaster aequalis)*
*36 (0–1) *vertical fascioles present between marginal pairs*
 78 (0–1) *distal amb–adamb articulation surface ada1 single transverse facet (paralleled in Odontaster validus)*
*105 (1–0) *distal bar of circumorals strongly angled to body of ossicle*
*106 (1–0) *distal bar of circumoral has double articular surface with oral*
 113 (0–1) *articulation structures between orals of a pair made of fine dentition*

The monophyly of the paxillosids receives weak support in this study (Bremer value of 1). Characters shared plesiomorphically with the outgroup, but not found in other neoasteroids, include the presence of an external spine-bearing face on the odontophore in many paxillosids, the upright, symmetrical nonimbricating construction of the ambulacral heads and the plesiomorphic nature of the ambulacral–adambulacral articulation, in which *ada3* is an interadambulacral articulation only. Additionally, the madreporite has a lateral position, as in Palaeozoic asteroids.

Apart from the Cribellina Fisher, 1911, which is a well-supported monophyletic group, other relationships within the Paxillosida have only weak Bremer support. It is interesting to note, however, that *Benthopecten* falls within the paxillosid clade as sister group to all other taxa. The order Notomyotida Ludwig, 1910 (see Blake 1987; Clark and Downey 1992) is taxonomically identical with the Benthopectinidae Verrill, 1899 and is not used here.

Clade CRIBELLINA Fisher, 1911

 3 (0–1) *3. interbrachial arcs curved, paralleled in Mediaster aequalis*
*36 (1–2) *cribriform organs present*

*72 (0–1) 72. *adambulacral–ambulacral articulation structure* ada1 *narrow, rugose*

*73 (0–1) *ambulacral–adambulacral articulations* ada2 *and* adada *confluent, form concavo-convex structure*

*86 (0–1) *mouth frame stellate in internal abactinal view*

*109 (0–1) *circumoral ossicles imbricate distally with proximal ambulacrals*

Taxa included here are *Ctenodiscus crispatus* and *Styracaster chuni*, in a well-supported monophyletic group (Bremer support of 4). The goniopectinid *Goniopecten demonstrans* Perrier, 1881 has synapomorphies 3, 36, 72 and 73.

Clade SURCULIFERA Gale, 1987

*10 (0–1) *tips of tube feet flat*

*13 (0–1) *stomach eversible with retractor system*

14 (0–1) *Tiedemann's pouches present (absent in* Forcipulatida*)*

17 (0–1) *madreporite central in position (MP >1) (reversed in* Brisinga costata*)*

*43 (0–1) *third ossicle integrated part of pedicellaria*

*46 (0–1) *internal adductor present on third ossicle of peds*

59 (0–1) *ambulacral heads imbricate proximally (reversed in many forcipulatids)*

*67 (1–0) *processes on ambulacrals for ambulacral–adambulacral muscles symmetrical*

71 (0–1) ada1 *comprises two facets (*ada1a, ada1b*) (reversed in* Odontaster validus*)*

*97 (0–1) *external face on odontophore entirely absent*

*114 (0–1) *apophysis of oral >25 per cent total height of ossicle*

This group includes all neoasteroids other than the paxillosids. Surculifera are characterized by diverse skeletal and soft-tissue synapomorphies and also developmental characters such as the presence of brachiolaria larvae and indirect development (Wada *et al.* 1996). The nature of the skeletal development immediately after metamorphosis is also distinctive, as five primary interradials and the centrale appear together after the terminals have formed and before radial and marginal elements appear (known in Asterinidae Gray, 1840, Solasteridae Viguier, 1878, Asteriidae Gray, 1840 and Echinasteridae Verrill, 1870; e.g. Komatsu *et al.* 1979). Developmental features are not included in the character list because the character states are not described for many species used in this study.

Clade TRIPEDICELLARIA nov.

15 (0–1) *pyloric complex large well developed (reversed in forcipulatids)*

90 (1–0) *oral ossicles project proximally to occupy much of oral ring (reversed in* Echinaster purpureus *and* Forcipulatida*)*

*104 (0–1) *odontophore-oral muscle insertion restricted to odontophore capsule*

113 (0–2) *articulation structures between orals of a pair (interradial surface) comprise discrete flat articulation surfaces, made of imperforate stereom (reversed in* Zoroaster fulgens, Heliaster helianthus *and* Brisingida*)*

*115 (0–1) *proximal circumoral bar of oral inclined proximally at 10–20 degrees*

118 (0–1) *oral with proximal blade-like extension (reversed in* Forcipulatida*)*

123 (0–1) *actinal surface of orals flattened (paralleled in* Calliasterella mira, *reversed in* Echinaster purpureus, Forcipulatida*)*

124 (0–1) *external (radial) surface of orals articulates with adjacent oral pair (reversed in* Forcipulatida*)*

127 (0–1) *oral spines form dense articulating cover over actinal face of orals (reversed in* Forcipulatida*)*

This group has reasonable Bremer support (2) and is named after homologies of pedicellarial structure identified between taxa assigned traditionally to the valvatids and forcipulatids (see homology section and Text-fig. 3, as well as Pls 3, 4). The identification of this group is one of the surprising results of the present study, as the Valvatida appeared to be monophyletic and well supported from most studies (e.g. Blake 1987). However, many of the characters of the Valvatida that were used to support the group are found to be plesiomorphic, and in the consensus cladogram (Text-figs 20, 21), the 'Valvatida' form a paraphyletic sister group to the Forcipulatida, plus *Echinaster purpureus*. *Asteropsis carinifera* is sister taxon to all other taxa of the group. *Archaster lorioli* and *Mediaster aequalis* form a poorly supported sister group to the remaining taxa. The core of the group, including *Nardoa variolata*, *Protoreaster nodosus*, *Acanthaster planci* and *Echinaster purpureus*, forms a paraphyletic stem lineage leading to the forcipulatids. This group has not been identified in any previous studies.

Unnamed clade; *Echinaster purpureus* + FORCIPULATIDA

11 (0–1) *tube feet suckered (paralleled in* Crossaster, Pteraster *and* Remaster*)*

18 (0–1) *madreporite fused with primary interradial and paired interradials (reversed in* Zoroaster*)*

*85 (0–1) *adambulacral spines very few (2–4)*

*88 (0–1) *proximal lateral process of odontophore projects into base of first podial opening*

*89 (0–1) *no lateral notch on odontophore, side of odontophore parallel with distal circumoral bar*

*91 (0–1) *adoral carina present*

*92 (0–1) *actinostome present*

93 (0–2) *odontophore trapezoidal in shape (paralleled in*
Benthopecten, Remaster, Pteraster, *reversed in*
Brisinga *and* Freyella)

94 (0–1) *odontophore abactinally convex, actinally concave*
odontophore trapezoidal in shape (paralleled in
Benthopecten, Remaster, Pteraster, *reversed in*
Brisinga *and* Freyella)

*99 (0–1) *odontophore articulates distally with circumoral*
ossicle

101 (0–1) *keel absent on odontophore (paralleled in* Benthopecten, Remaster, Pteraster)

*122 (0–1) *actinal surface of orals small, inconspicuous*

*128 (0–1) *oral spines very few*

The discovery that *Echinaster purpureus* shares numerous synapomorphies with the Forcipulatida is unexpected, because the echinasterids have been placed in the Spinulosida since inception of the group (Perrier 1894). Indeed, Blake (1987, fig. 10) retained the order Spinulosida solely for the Echinasteridae and transferred all other families to the Velatida, which he identified as sister group to the Spinulosida, both included in a superorder Spinulosacea. The Spinulosacea were not well supported, being characterized mostly by embryological and larval characters that are known in very few taxa, and the group was not identified in the present study. However, Viguier (1878) had already commented upon similarities between the mouth frame of *Echinaster* Müller and Troschel, 1840 and *Asterias* Linnaeus, 1758, noting particularly the morphology of the odontophore. The inferred relationship in the present study is based largely upon similarities in mouth frame development which appear to be robust. From Viguier's (1878) excellent figures of the mouth frame, it appears that *Mithrodia clavigera* (Lamarck, 1816) (family Mithrodiidae Viguier, 1878) is closely related to *Echinaster* and probably belongs to the same clade.

Clade FORCIPULATIDA Perrier, 1884

*87 (0–1) *odontophore shallow in position, just beneath*
circumoral heads

*107 (0–1) *circumoral ossicles longer than tall*

*111 (0–1) *circumoral heads elongated (=3–6 ambs)*

*116 (0–1) *abactinal interradial interoral muscle* (abiim)
attachment deep, set on proximal circumoral bar

*119 (0–1) *groove for ring nerve on oral shallow, superficial*

*120 (0–1) *articulation between first adamb and oral set at*
70–90 degrees to interoral surface

*121 (0–1) *first adambulacrals enlarged, proximal articular*
surface broad, form extension to orals

The Forcipulatida are a very well-supported group (Bremer support >8), although a large number of the synapomorphies traditionally used to characterize the group are

actually shared with *Echinaster* and therefore transfer to the previous unnamed clade. The phylogeny given here (Text-fig. 21) fails to resolve relationships between *Zoroaster*, the Brisingida and a clade including *Asterias* and *Heliaster* Gray, 1840. The phylogeny presented by Mah (2000) included many more taxa (25) and placed a paraphyletic Pedicellasteridae Perrier, 1884 as basal to all other forcipulatids. The most derived group included the Brisingida Fisher, 1928, Labidiasteridae Verrill, 1914, Pycnopodiinae Fisher, 1928, Coscinasteriinae Fisher, 1923 and some Asteriidae.

Clade BRISINGIDA Fisher, 1928

16 (1–2) *abactinals reduced to small ossicles set in*
membrane (convergent with Benthopecten)

*56 (0–1) *abactinal transverse ambulacral muscle deeply*
inset between regions of dentition

*57 (0–1) *dentition separated into proximal and distal*
parts

*59 (1–3) *amb head upright, vertebra-like*

*62 (0–1) *longitudinal interambulacral articulation*
elongated abactinally–actinally

*63 (0–1) *longitudinal interambulacral articulation strongly*
concavo-convex

*83 (0–1) *distal abradial extension on adambulacrals*

*86 (2–3) *mouth frame ring-like, inflexible*

*87 (2–3) *odontophore flush with circumoral heads*

*93 (0–2) *odontophore block-like, rectangular*

*98 (1–2) *odontophore fused with oral ossicles*

*100 (0–1) *odontophore forms integral part of wall of first*
podial opening

*108 (0–1) *proximal oral-circumoral dentition flat, broad*

*110 (0–1) *transverse intercircumoral muscle internal*

*112 (0–1) *circumoral dentition forms immobile contact*

The Brisingida are a strongly supported monophyletic group nested within the Forcipulatida (Bremer support of >8) with numerous synapomorphies of the ambulacral groove and mouth frame (see also Mah 2000).

Clade SPINULOSIDA Perrier, 1884

5 (0–1) *shafts of spines composed of coarse, elongated*
glassy trabeculae (convergent with Brisinga *and*
Freyella)

32 (1–0) *marginals not block-like (homoplastic)*

38 (0–1) *actinals set in transverse rows running from*
adambulacrals to marginals (convergent with
paxillosids Ctenodiscus, Radiaster, Luidia,
Astropecten)

59 (1–2) *ambulacral head with strong proximal imbrication*
(homoplasy with A. planci)

62 (1–2) *interadambulacral articulation large, diffuse*

65 (0–1) *ambulacral shafts composed of glassy, transversely elongated trabeculae (convergent with* Heliaster *and* Asterias)

77 (0–1) *interadamb articulation adada absent*

79 (0–1) *amb–adamb articulations ada2 and ada3 united to form crescentic or hourglass-shaped facet*

93 (0–1) *odontophore short, broad with transverse bar (reversed in* Pteraster, Remaster)

117 (0–1) *vertically elongated groove for ring vessel on apophyse of oral (reversed in* Crossaster papposus)

*125 (0–1) *actinal part of oral made up of rounded lateral rim and central mound*

*126 (0–1) *discrete oral and suboral spines*

128 (0–1) *oral spines few*

As originally designated, Perrier's Spinulosa comprised the Echinasteridae, Asterinidae, Poraniidae Perrier, 1875, Pterasteridae Perrier, 1875 and Solasteridae. This use of the name was continued by Fisher (1911) and by Spencer and Wright (1966), who amended it to Spinulosida. Blake (1984) transferred the Asterinidae and Poraniidae to the Valvatida and later (Blake 1987) resurrected Perrier's Velatida for much of the remainder of the Spinulosida, restricting the latter order to the Echinasteridae. This course of action was followed by Clark and Downey (1992).

The present study supports the return to a traditional concept of the Spinulosida close to that used by Fisher and Spencer and Wright, with two important differences. First, *Odontaster* Verrill, 1880, traditionally placed in the Valvatida, shares most of the skeletal synapomorphies of other spinulosids. Its placement in the valvatids has perhaps been guided by the possession of large marginal ossicles in some species that therefore superficially resemble goniasterids. Second, *Echinaster* falls close to the forcipulatids in the consensus tree and is therefore removed from the group. As defined herein, the spinulosids are well characterized by ambulacral and adambulacral structure, spine morphology and construction and spination of the oral ossicles.

Although the transfer of the Asterinidae from the Valvatida to the Spinulosida conflicts with views expressed by Blake (1981, 1987) and Clark and Downey (1992), it receives strong support from the molecular work of Matsubara *et al.* (2004), who found that mitochondrial and nuclear 18S rDNA data strongly supported a close relationship between the Solasteridae and Asterinidae.

Clade VELATIDA Perrier, 1894

*8 (0–1) *interradial chevron plates present*

*9 (0–1) *interradial grooves present*

*29 (0–1) *spines on abactinal ossicles webbed*

*60 (1–2) *ambulacral head with distal flange to carry additional longitudinal interambulacral muscle*

*80 (0–1) *adambulacral extensions present*

*81 (0–1) *abradial muscle present between successive adambulacral extensions*

*82 (0–1) *abradial articulation surface between abradial extensions*

93 (1–2) *odontophore rhomboidal–trapezoidal (convergent with* Benthopecten, Echinaster, Forcipulatida)

94 (0–1) *odontophore abactinally convex, actinally concave (convergent with* Benthopecten, Echinaster, Forcipulatida)

*103 (0–1) *specialized articulation surface for chevron plates on abactinal distal odontophore surface.*

The Velata of Perrier (1894) was erected to include the distinctive Pterasteridae, but probably fell into disuse because most workers simply included that family in the Spinulosida. Blake (1987) subsequently resurrected the group as the Velatida and used the name to include most of what had previously been included in the Spinulosida (e.g. Spencer and Wright 1966), leaving only the Echinasteridae in that group (see above). This classification was followed by Clark and Downey (1992). Re-investigation of the Korethrasteridae Danielssen and Koren, 1884 and Pterasteridae has revealed the presence of numerous distinctive skeletal synapomorphies, including the interradial chevron plates and adambulacral extensions, and the Velatida is redefined as a strongly supported group (Bremer values of 5) and includes the genera *Korethraster* Wyville Thomson, 1873, *Remaster* Perrier, 1894, *Peribolaster* Sladen, 1889 and the families Pterasteridae and Myxasteridae Perrier, 1885.

THE FOSSIL RECORD OF THE NEOASTEROIDS

Fossil record and stratigraphical distribution of extant neoasteroid families

Comparison of the stratigraphical distribution of a group with the inferred order of appearance of constituent taxa on a cladogram provides a valuable means of assessing the probable accuracy of any given phylogeny (Smith 1994). Because the neoasteroid taxa considered herein are all modern species with no fossil record, the distribution of their families is taken as a proxy. A problem here is the probable paraphyly or polyphyly of some asteroid families (e.g. Goniasteridae (see Blake 1987); Asterinidae, Asteropseidae Hotchkiss and Clark, 1976 (see O'Loughlin and Waters 2004); Korethrasteridae) and incorrect assignation of often fragmentary or poorly preserved fossil material to extant families. Clearly, the first and last occurrences of nonmonophyletic taxa have no phylogenetic significance (Patterson and Smith 1987). Both are considered in the discussion of successive taxa. Even if some doubt surrounds the precise taxonomic assignment of these fossils, the first occurrence of important characters (e.g. alveolar, straight and crossed pedicellariae; adambulacral extensions, cribriform organs) assists the dating of branches and nodes of the consensus tree (Text-figs 20, 21).

Luidiidae – Luidia *cf.* ciliaris: *Miocene.* The earliest records of this monogeneric family are of *Luidia* aff. *ciliaris* and of a similar taxon from the lower and lower middle Miocene of south-west France (Aquitaine), Portugal, the Netherlands and Poland (Kaczmarska 1987; Jagt 1991; Pereira *et al.* 2003; J. W. M. Jagt, pers. comm. January 2010). These records belongs to the *Ciliaris* Group of Döderlein (1920), indicating that significant divergence within the genus had taken place by the Early Miocene.

Radiasteridae – *no fossil record.* The Radiasteridae is monogeneric, and the four constituent species appear to be closely related. The only fossil assigned to this family is *Betelgeusia reidi* Blake and Reid, 1998, from the upper Albian or lower Cenomanian of Texas. Examination of the material of this species indicates that it is a typical astropectinid with broad interradial arcs, morphologically unique, but with affinities to the extant *Dipsacaster* Alcock, 1893. The evidence for this can be summarized as follows (Text-fig. 22):

– The marginals in *Betelgeusia reidi* are proportionately large, transversely broad and short with raised articulation surfaces and fascioles between successive plates, as in many astropectinids (see Blake 1973, pl. 15, figs 6–11; compare with Blake and Reid 1998, fig. 8/3–4, 6–7, 14–15; Text-fig. 22A–D herein). The inferomarginals of *Betelgeusia reidi*, as in most astropectinids, carry both smaller and larger spines arranged in a transverse row; the superomarginals have a covering of even, shorter spines. In contrast, the marginals in *Radiaster tizardi* are small, elongated and paxilliform (e.g. Clark and Downey 1992, fig. 17a; Text-fig. 22E, F herein), and supero- and inferomarginals carry small tufts of similar spines.

– The ambulacrals of *Betelgeusia reidi* are typically astropectinid (compared with Blake 1973, pls 14–15, with strongly asymmetrical proximal and distal ambulacral–adambulacral muscle attachment sites. The ambulacrals in *Radiaster tizardi* are more like those in *Luidia* with nearly symmetrical muscles.

Astropectinidae – Pentasteria? liasica *Villier, Kutscher and Mah, 2004: Lower Jurassic, Toarcian.* This family includes some 25 morphologically diverse, extant genera, many of which are of infrequent occurrence in deeper water (Clark and Downey 1992). There has never been a cladistic assessment of the family, and it cannot be assumed that it

is monophyletic. Jurassic taxa with astropectinid-like body plans and marginal morphologies (intermarginal fascioles, large bladed inferomarginal spines) are numerous and have been mostly assigned to *Pentasteria* Valette, 1929 (see Hess 1955).

However, Blake (1986) referred well-preserved material of *Pentasteria* (*Archastropecten*) *portlandicus* Hess, 1955 from the Tithonian of the UK to the genus *Pseudarchaster* Sladen, 1889. *Pseudarchaster* is usually placed in the family Goniasteridae (see Blake 1987) and has flat-tipped tube feet, an anus and a brachiolarian larva (Wada *et al.* 1996), but resembles astropectinids in body form and some details of ossicular morphology (e.g. it has shallow intermarginal fascioles). This assignation is significant, because *P.* (*A.*) *portlandicus* is a typical member of this common and widely distributed Jurassic genus with implications for the entire Jurassic record of astropectinids if it is correctly referred to *Pseudarchaster*. A re-examination of the type (NHM E 13739) of *Pentasteria* (*Archastropecten*) *portlandicus* has revealed close similarities of marginal spination to extant astropectinids, such as *Proserpinaster* Fell, 1963 (Text-fig. 23A, B). In both, the inferomarginals possess a distal fringe of elongated, blade-like spines that become larger towards the lateral margin and a covering of short, proximally imbricating, scale-like, striated spines over the remainder of the ossicle (see also Fisher 1919*a*). In contrast, the inferomarginal spination in *Pseudarchaster* (Text-fig. 23C) comprises a tesselation of flat, rather widely spaced, polygonal spines and a transverse row of four or five flattened, short-bladed spines along the distal margin.

Examination of superbly preserved examples of Jurassic starfish such as *Pentasteria* (*Archastropecten*) *longispina* Hess, 1968 (see Hess 1987, fig. 2) from the Oxfordian of Switzerland is convincing of the astropectinid affinity of this species. The actinal surface figured by Hess shows oral ossicles, actinal fasciolar channels, adambulacrals and inferomarginals, which are closely similar to those of modern astropectinids, but which do not fit into any living genus.

The oldest Jurassic asteroid referred to the Astropectinidae is *Pentasteria* (*Archastropecten*)? *hastingiae* (Forbes, 1848) from the middle Lias of Yorkshire, UK (Hess 1955). However, this species has small marginal ossicles and enlarged, flat abactinal ossicles, quite unlike the paxillae seen in true astropectinids, and it must be treated as *incertae sedis*. The oldest record is therefore *Pentasteria? liasica* from the Toarcian of western France (Villier *et al.* 2004*a*).

Goniopectinidae – Chrispaulia jurassica *sp. nov.; Upper Jurassic, Oxfordian.* The Ctenodiscinae and Goniopectininae are here treated as subfamilies within the Goniopec-

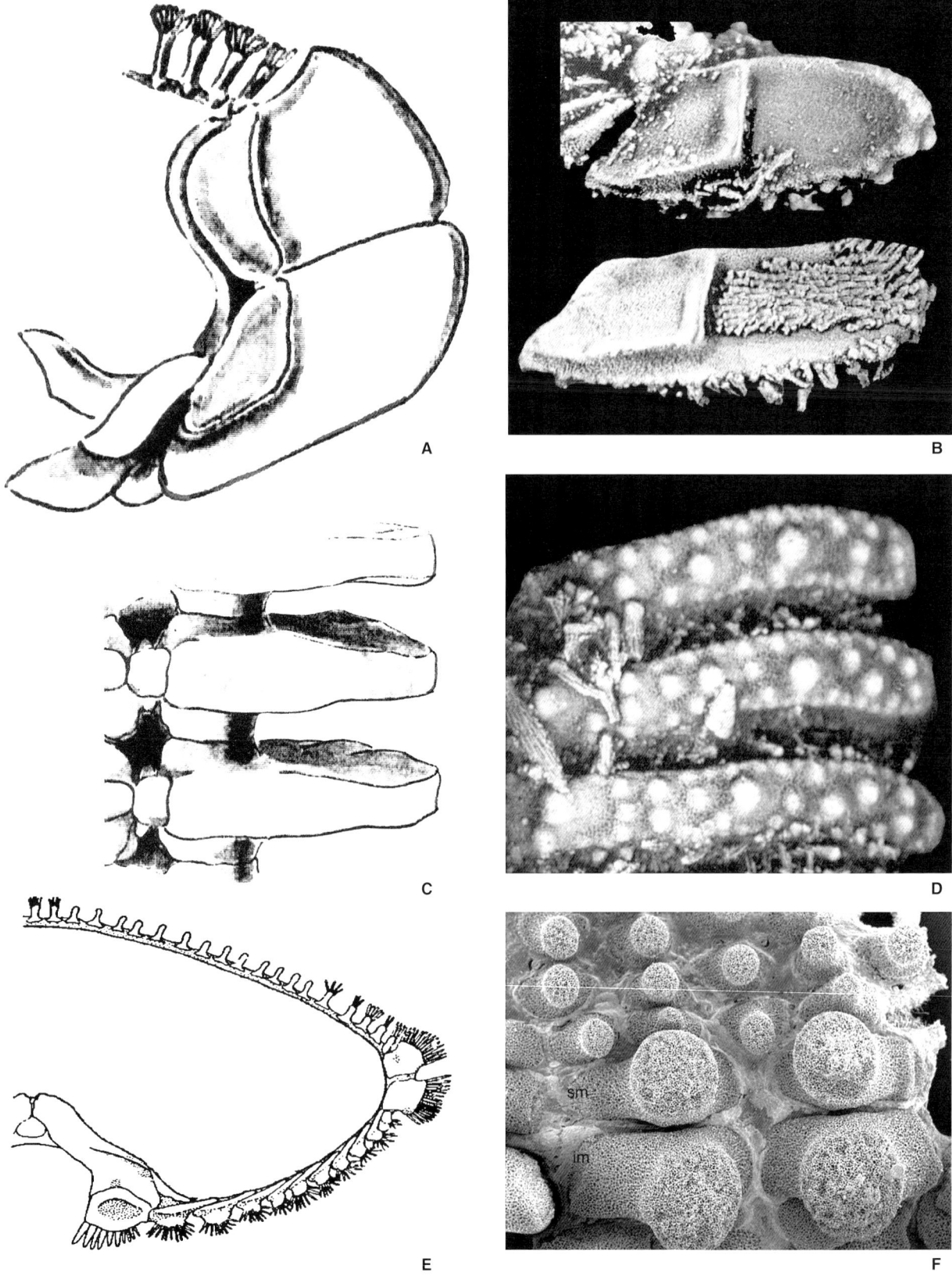

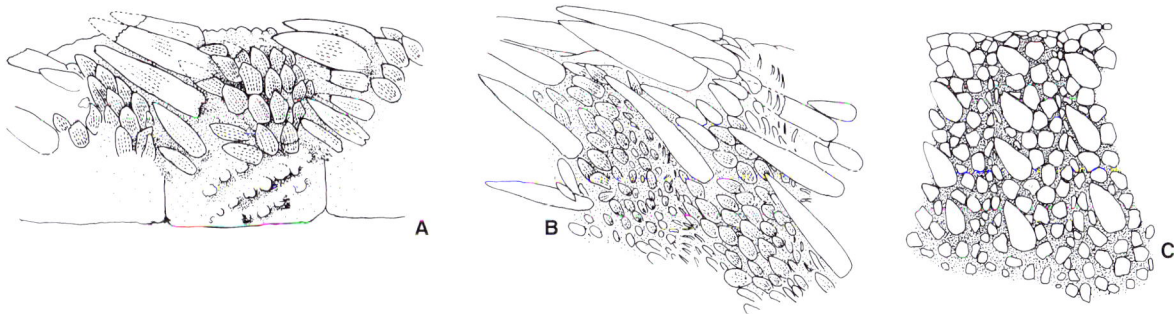

TEXT-FIG. 23. Inferomarginal spination in *Pentasteria* (*Archastropecten*) *portlandensis* (A) (Tithonian, Portland Limestone, UK; NHM E13739, original of Hess 1955, pl. 4, figs 4–6), *Proserpinaster neozelandicus* (Mortensen, 1925) (B) and *Pseudarchaster parelli* (Düben and Koren, 1846) (C; see Table 1 for locality details). A covering of short, striated, scale-like spines, and larger bladed spines on the distal/lateral part of the ossicle is very similar in *P.* (*A.*) *portlandensis* and *Proserpinaster neozelandicus* and typical of a group of astropectinids also including the extant *Persephonaster* and *Astromesites* Fisher, 1913 (see Fisher 1919a, pls 20–28). In contrast, typical *Pseudarchaster* spination (C) comprises short, polygonal smooth spines, and a distal transverse row of 3–4 larger, short, flattened, oval spines. *Pentasteria* (*Archastropecten*) *portlandensis* is an astropectinid, not a *Pseudarchaster*. A, ×12; B, ×10; C, ×4.

tinidae, following Clark and Downey (1992). The only proposed fossil ctenodiscine, *Paleoctenodiscus campaniurensis* Blake, 1988a, was transferred to the Porcellanasteridae Sladen, 1883 by Gale (2005). A new goniopectinid from the lower Oxfordian of Switzerland and France, *Chrispaulia jurassica* sp. nov., is described below (section 'Systematic Palaeontology').

Porcellanasteridae – *Paleoctenodiscus campaniurus*; *Upper Cretaceous, Campanian.* The only fossil porcellanasterid is *Paleoctenodiscus campaniurus* from the Campanian of Mexico (Gale 2005).

Benthopectinidae – Jurapecten hessi *gen. et sp. nov.; Upper Jurassic, Oxfordian.* There have been numerous claims of fossil benthopectinids, from the Hettangian (Lower Jurassic) onwards (e.g. Spencer and Wright 1966; Blake 1984; Blake and Reid 1998). However, the majority of these share only very superficial characters with the group, such as elongated marginal and abactinal spines and lack the numerous distinctive skeletal synapomorphies of the family originally described by Blake (1972). The Palaeobenthopectininae Blake, 1984 is a heterogeneous assemblage of spinulosids ('velatids', Villier *et al.* 2009) and forcipulatids. The earliest fossil that displays the distinctive skeletal synapomorphies of the Benthopectinidae is *Jurapecten hessi* gen. et sp. nov. from the Oxfordian of France and

Switzerland (see below, section 'Systematic Palaeontology').

Goniasteridae – Comptoniaster vrinensis; *Lower Jurassic, Toarcian.* This is a large and diverse family, united by plesiomorphic characters such as the large marginals. Blake (1987) demonstrated that it is paraphyletic with one subfamily, the Pseudarchasterinae, falling as sister group to the other valvatids. Nevertheless, Clark and Downey (1992) included *Pseudarchaster* in the goniasterids. The earliest well-preserved asteroid referable to the family is *Tylasteria berthandi* (Wright, 1880) from the Middle Jurassic (Bajocian) of Switzerland (Hess 1972). Villier *et al.* (2004a) recorded marginal ossicles of *Comptoniaster vrinensis* from the Toarcian (Lower Jurassic) of France and Germany, which constitute the earliest record of the family.

Archasteridae – Archaster patersoni; *Miocene. Archaster patersoni* Spencer, 1915, from the Miocene of Port Elizabeth, South Africa, is the only fossil record of this family.

Ophidiasteridae – Denebia americana; *Middle Cretaceous, Cenomanian.* This large and diverse modern family is rather poorly characterized morphologically and is separated from the closely related oreasterids largely on its possession of narrow rather than broad arms. *Denebia*

TEXT-FIG. 22. Affinities of *Betelgeusia reidi* Blake, 1998 from the Middle Cretaceous (Albian) of Texas. B, D, inferomarginals in proximal or distal aspect (B) and actinal aspect (D), after Blake and Reid (1998, fig. 8/3; SMU 30049; fig. 8/4, SMU 30050; fig. 8/14, SMU 30042), compared with those (A, C) of the extant astropectinid *Tethyaster aulophora* (Fisher 1919a, pl. 39, fig. 1a, d). Note the short, very broad inferomarginals with large fasciolar surfaces and proportionately small articulation surfaces for the adjacent IMs. E, cross-section of arm of *R. tizardi* after Clark and Downey (1992, fig. 17A). F, SEM image of mid-arm of *Radiaster tizardi* (Table 1) showing small elongated paxilliform development of marginals. Note important differences between marginal development of astropectinids and *Radiaster tizardi*; *Betelgeusia reidi* is an astropectinid. A–D, ×10. E, ×7. F, ×30.

americana (Adkins, 1928) and *Altaria wintoni* (Adkins, 1928) from the Middle Cretaceous (Albian–Cenomanian) of Texas appear to be the earliest records (Blake and Reid 1998).

Oreasteridae – Goniodiscaster?; *Eocene.* Although Mesozoic fossil taxa were referred to *Pentaceros* Gray, 1840 (synonym of *Oreaster* Müller and Troschel, 1842) in the nineteenth century, many of these have subsequently been placed in the predominantly Jurassic–Cretaceous family Stauranderasteridae Spencer, 1913. The taxonomy and phylogeny of this family have been reviewed by Villier *et al.* (2004c), who concluded that it constituted a monophyletic group (Jurassic–Eocene), related to, but separate from, the Oreasteridae Fisher, 1911 (Eocene–Recent). However, stauranderasterids and oreasterids appear to be closely related, and it is difficult to identify any synapomorphies which are unique to stauranderasterids. Cretaceous stauranderasterids possess cupules, which would place them within the clade including Oreasteridae, Ophidiasteridae Verrill, 1870 and Acanthasteridae Sladen, 1889. It is concluded provisionally that stauranderasterids are stem group oreasterids. However, the first definitive fossil oreasterid is *Goniodiscaster*(?) from the Eocene of Mexico (Blake 1986, figs 3–8).

Acanthasteridae – Acanthaster planci; *Holocene.* The only known fossil acanthasterid is *A. planci* from the Holocene of north-east Australia (Walbran *et al.* 1989).

Echinasteridae – no fossil record. There have been various claims of fossil echinasterids in the literature, from *Henricia*? *venturana* Durham and Roberts, 1948 from the Cretaceous of California (an indeterminate asteroid), *Protothyraster priscus* Hess, 1970 from the Hauterivian of Switzerland (a possible forcipulatid) and *Echinaster jacobseni* Rasmussen, 1972 from the Eocene of Denmark (a poraniid; see below). Re-examination of *Protothyraster priscus* shows that it has forcipulatid abactinal plating (see below; section 'Systematic Palaeontology', under *Terminaster cancriformis* – Remarks), with inset adradial ossicles and a strong, raised row of radials and is provisionally assigned to Forcipulatida *incertae sedis*. *Echinaster jacobseni* displays various characters of the Poraniidae including the deep sites for insertion of the interadambulacral ossicles and the form of the oral ossicles.

Asteriidae – Germanasterias amplipapularia; *Lower Jurassic, Hettangian.* The earliest taxa referred to the Asteriidae in the literature are *Germanasterias amplipapularia* Blake, 1990 and *Hystrixasterias hettangiurnus* Blake, 1990 from the Lower Jurassic (Hettangian) of Germany. These agree with asteriids in overall construction and some details of ossicular morphology (e.g. short, compressed ambulacrals,

oral and circumoral morphology), but possess an unusual form of straight pedicellariae which are very broad, flattened in the plane of the valves and have a sharp, narrow tip (Text-fig. 24D–G). These are enormous (up to 5 mm in length), concentrated around the ambitus and entirely unknown in extant asteriids or other forcipulatids, in which the straight pedicellariae are either felipedal or lanceolate (Text-fig. 24H–O). Additionally, the author has failed to find crossed pedicellaria on either Hettangian genus, whereas all extant asteriids have both straight and crossed pedicellariae. Crossed pedicellariae are identified in *Dermaster boehmi* de Loriol, 1899 (see Hess 1972) from the Bajocian. The Hettangian record from Germany is provisionally taken as the earliest asteriid because the overall morphology of the genera is comparable with that of the Asteriidae, but should probably be placed in a separate family.

Heliasteridae – Heliaster microbrachius; *Pliocene.* The only fossil record of this monogeneric family is the occurrence of *Heliaster microbrachius* Xantus, 1860 from the Pliocene of Florida (Jones and Portell 1988).

Freyellidae. No fossil record.

Brisingidae – Brisingella *sp.; Miocene.* The earliest, and only, fossil record of brisingids is that of *Brisingella* sp. from the lower Miocene (Burdigalian) of Chita Peninsula, Japan (Masatoshi 1987).

Zoroasteridae – Zoroaster *aff.* fulgens; *Eocene.* Hess (1974) referred *Terminaster cancriformis* (Quenstedt, 1876) from the Callovian–Oxfordian (Upper Jurassic) of the United Kingdom and Switzerland to this family based upon a similarity of body form and development of external ossicles (abactinals, marginals), supported by Mah (2007). New material of *Terminaster cancriformis* described herein (see section 'Systematic Palaeontology') has justified the placement of the genus in a new family, Terminasteridae. *Zoroaster* aff. *fulgens* from the Antarctic Peninsula is the oldest confirmed record of a zoroasterid (Blake and Zinsmeister 1979).

Asteropseidae – Diclidaster gevreyi; *Lower Jurassic, Hettangian.* The Asteropseidae is a poorly diagnosed small family (Clark and Downey 1992) in which were included the genera *Asteropsis* Müller and Troschel, 1840, *Dermasterias* Perrier, 1875, *Valvaster* Perrier, 1875, *Petricia* Gray, 1847 and *Poraniella* Verrill, 1914 (Hotchkiss and Clark 1976; Clark 1984; Clark and Downey 1992). Subsequently, O'Loughlin and Waters (2004) transferred *Dermasterias* to the Asterinidae on molecular evidence; the monophyly of the remainder of the family must be considered uncertain. Spencer and Wright (1966) assigned *Diclidas-*

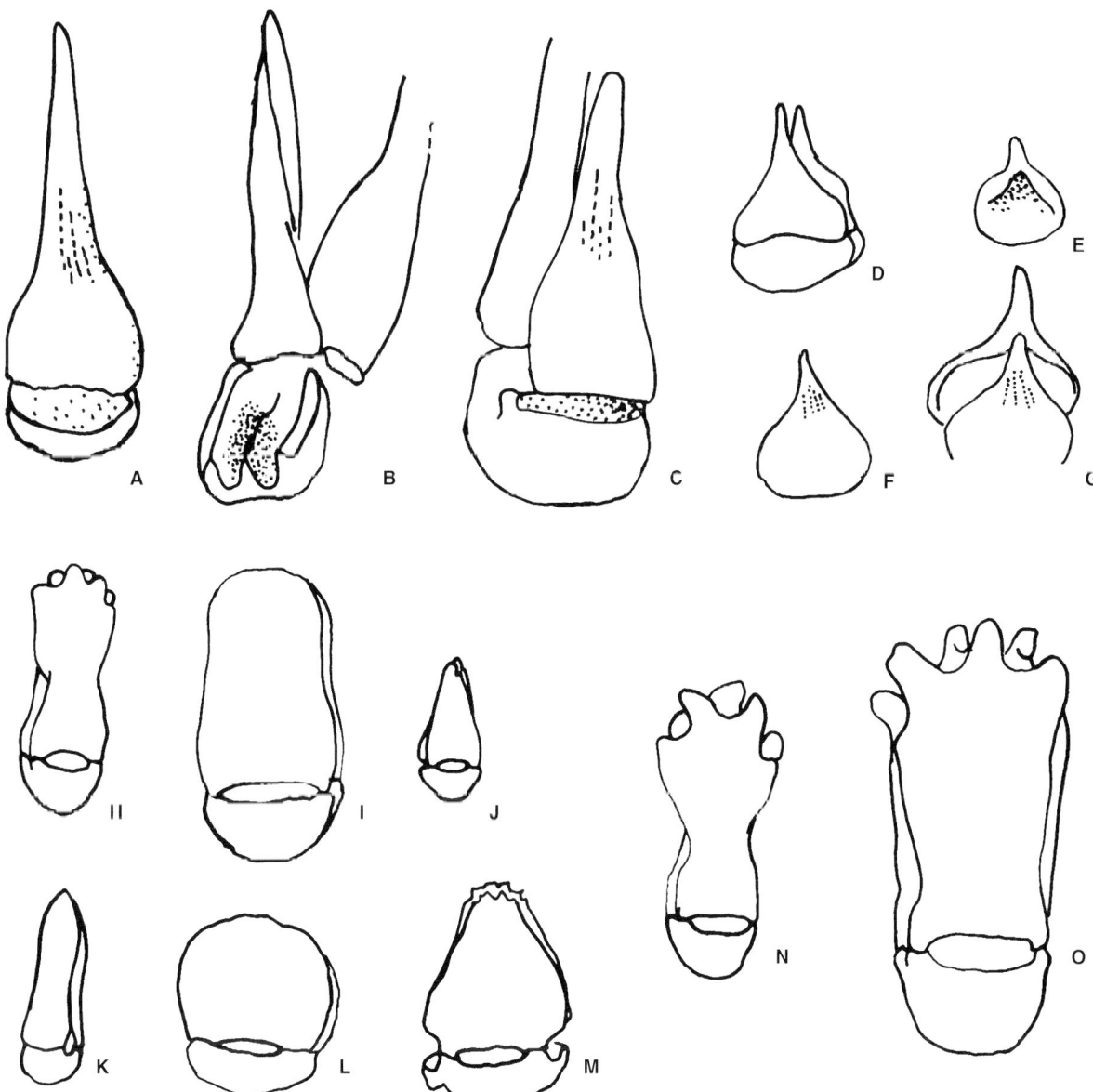

TEXT-FIG. 24. Comparison of pedicellariae of Recent Asteriidae and Jurassic forcipulatids. A–C, interradial pedicellariae of 'Asterias' *gaveyi* Forbes, 1850, holotype, Mickleham Tunnel, Chipping Camden, Gloucestershire, UK, Lower Jurassic (Pliensbachian, *capricornu* Zone), NHM E 1638. D–G, pedicellariae from the ambitus of *Hystrixasterias hettangiurnus* Blake, 1990, paratype, Naturhistorisches Museum Basel, M 9688, Lower Jurassic (Hettangian, *angulata* Zone), Württemberg, Germany. H–O, straight pedicellariae of extant asteriids and pycnopodiids, from the north-east Pacific, after Fisher (1928, 1930). H, K, *Orthasterias koehleri* (de Loriol, 1897b), Fisher, 1928 (pl. 66, fig. 5, fig. 8a). I, *Pycnopodia helianthoides* (Brandt, 1835). J, L, *Leptasterias alaskensis asiatica* Fisher, (1930, pl. 48, figs 1a, 2). O, *Lethasterias nanimensis nanimensis* Fisher, 1928 (pl. 61, fig. 2c). M, *Pisaster giganteus* Fisher, 1928 (1930, pl. 77, fig 6b). N, *Leptasterias arctica* (Murdoch, 1885), Fisher (1930, pl. 9, fig. 8). All Early Jurassic forcipulatids with skeletal constructions close to extant Asteriidae possess only straight pedicellariae of the constructional type shown in figures A–G, in which the valves taper sharply to a point. These are different to the straight pedicellariae of extant asteriids, which are dominantly of felipedal type (H, N), or broad, lanceolate form (J–M). A–G, ×12; H–O, ×25.

ter *gevreyi* de Loriol, 1897a from the Hettangian of the Ardèche, south-east France, to the Valvasteridae Viguier, 1878, a family later synonymized with the Asteropseidae by Blake (1987). The whereabouts of the single type specimen are unknown, but de Loriol's good illustration allows assessment of the taxon. Three rows of large abactinals and superomarginals are present in the arm, and large tongue-like, bivalved alveolar pedicellariae are attached to abactinals and marginals. The taxon is apparently closer to *Valvaster* than any other extant asteroid, and the specimen is important in any case as it demonstrates that alveolar pedicellariae were present in the earliest Jurassic.

Odontasteridae – Odontaster priscus; *Middle Jurassic, Bajocian.* The four genera recognized (Clark and Downey 1992) are united by synapomorphies including the presence of distinctive, hyaline-tipped, distally directed, unpaired oral spines, and the family is likely to be monophyletic. A single record of fossil odontasterid, from the Middle Jurassic (Bajocian) of New Zealand, *Odontaster priscus* Fell, 1954, is based on a curiously preserved specimen, which was interpreted by Fell as a cast of the body cavity. The sole character upon which affinity is based is comparison of the grossly enlarged, rounded circumoral heads, with concave insertion sites for the abactinal transverse circumoral muscles and coarse dentition, also found in *Odontaster meridionalis* (E.A. Smith, 1876) and *Odontaster crassus* Fisher, 1905 (see Fisher, 1911). Fell (1954) stated that, 'new material shows that this is not a true *Odontaster*'. The difficulty here is that the shape and size of the circumoral heads is known in so few asteroids that it cannot be certain that this is a characteristic of the odontasterids alone. For example, similar structures are present in the astropectinid *Pentasteria* (*Archastropecten*) *cottesswoldiae* (Buckman, 1844), figured by Wright (1862, pl. 9, fig. 1). The record is provisionally accepted.

Poraniidae – Sphaeriaster jurassicus; *Middle Jurassic, Bajocian.* The Poraniidae is a well-defined modern family (Clark 1984; Clark and Downey 1992), which is additionally characterized by numerous features of the ambulacral groove and mouth frame ossicles and probably is monophyletic. There is only one recorded fossil poraniid, *Sphaeriaster jurassicus* Hess, 1972 from the Bajocian of Switzerland, based upon two moderately well-preserved individuals. Assignation to the family is based upon overall shape similarity and the almost complete reduction of the extraxial skeleton, a feature characteristic of various poraniid genera (see Clark and Downey 1992). Unfortunately, the material of *Sphaeriaster jurassicus* is insufficiently well preserved to enable identification of skeletal features diagnostic of the family. As other asteroids have also dramatically reduced the extraxial skeleton, such as the recently described oreasterid *Astrosarkus* Mah, 2003 and the pycnopodiid *Lysastrosoma* Fisher, 1928, some doubt must attend the assignation of *S. jurassicus* to the Poraniidae. The only other fossil poraniid of which the author is aware is '*Echinaster*' *jacobseni* from the Eocene of Denmark (see below).

Asterinidae – Mesotremaster felli; *Middle Jurassic, Bajocian.* The Asterinidae is the only asteroid family for which there is a molecular phylogeny (O'Loughlin and Waters 2004). This unfortunately does not include the rare, deep-water *Tremaster mirabilis* Verrill, 1880, which nevertheless is provisionally included in the family, in spite of the reservations of O'Loughlin and Waters (2004). The Middle Jurassic *Mesotremaster felli* Hess, 1972 from the Bajocian of Switzerland is closely related to *Tremaster mirabilis* and is therefore identified as the earliest asterinid.

Solasteridae – Plesiosolaster moretonensis; *Middle Jurassic, Bajocian.* The solasterids constitute a small and well-characterized family with nine constituent genera (Clark and Downey 1992) and are likely to be monophyletic. The first fossil occurrence is *Plesiosolaster moretonensis* (Forbes, 1856) from the Bajocian of Switzerland (Hess 1972).

'*Korethrasteridae*' (=stem group Velatida) – Protremaster uniserialis; *Lower Jurassic, Hettangian.* The korethrasterid genera are paraphyletic sister taxa forming the stem group of the Velatida and as such the family has no phylogenetic validity. However, the first appearance of an asteroid with a velatidan character, the adambulacral extensions, is *Protremaster uniserialis* Smith and Tranter, 1985 from the Hettangian of Antarctica. This species displays a number of close morphological similarities with the extant *Korethraster hispidus* Wyville Thomson, 1873, including the scale-like imbricate form of the abactinal ossicles (Text-fig. 25A, D, E), and the short, very broad adambulacral ossicles (Text-fig. 25B, C).

Pterasteridae – Savignaster wardi gen. et sp. nov.; *Upper Jurassic, Oxfordian.* Pterasterids constitute a well-characterized group of asteroids for which a provisional morphological phylogeny has recently been published (Villier *et al.* 2004b). A new basal pterasterid, *Savignaster wardi* gen. et sp. nov., from the upper Oxfordian of the Swiss and French Jura is described below (see section 'Systematic Palaeontology').

PERMO–TRIASSIC ASTEROIDS: ORIGIN AND EARLY RADIATION OF THE NEOASTEROIDS

Permian asteroids are well known from the Artinskian of Australia, which has yielded six genera, two of which are still undescribed. *Permaster grandis* (see Kesling 1969) possesses some of the neoasteroid synapomorphies listed here such as the reduced external face on the odontophore and the presence of transverse ambulacral muscles. However, it lacks other characters possessed by all post-Palaeozoic asteroids, such as separation of the odontophore from the marginals, development of actinal os-

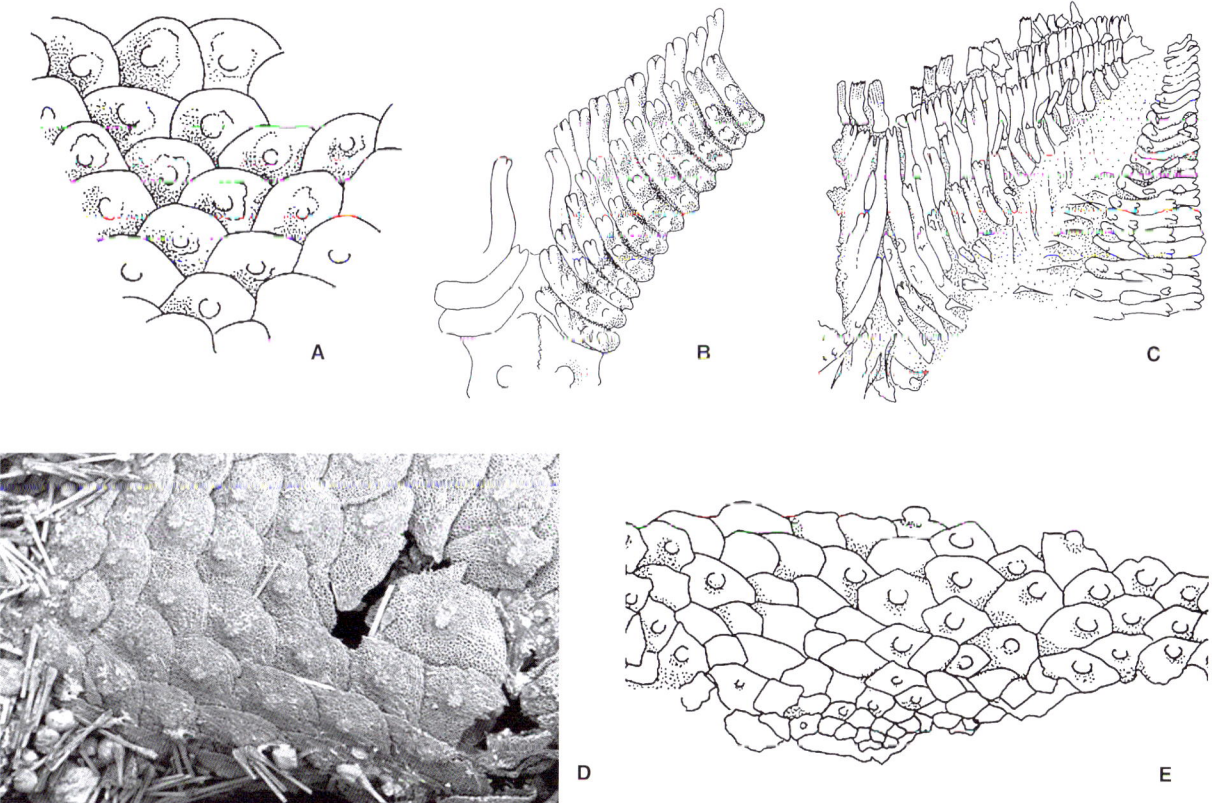

TEXT-FIG. 25. Comparison of skeletal construction in extant *Korethraster hispidus* Wyville Thomson, Faroe Islands (Table 1) (A, B, D) and *Protremaster uniserialis*, unique specimen from the Lower Jurassic (Sinemurian), Alexander Island, Antarctic Peninsula, BAS KG 2986.17 (C, E). A, C, *Korethraster hispidus*, enlargement of abactinal ossicles. B, actinal aspect of interradius, denuded of spines. *Protremaster uniserialis*, C, actinal view of interradius, E, abactinal surface. Note similarities of abactinal and adambulacral construction. However, in *Protremaster uniserialis*, the interradial grooves are closed by abactinal ossicles, and a calcified duct opens on the actinal and abactinal surfaces, as in *Tremaster mirabilis* Verrill, 1880. In *Korethraster hispidus*, the grooves are open like those of *Remaster gourdoni* (see Text-fig. 2A, B). A, D, ×20; B, ×10; C, E, ×5.

sicles and presence of superomarginals. The first probable neoasteroid appeared in the Middle Triassic, upper Anisian (Blake and Hagdorn 2003), and there is thus a gap of about 40 million years between the Artinskian and upper Anisian from which not a single specimen of asteroid has been described. The catastrophic extinctions at the P-T boundary were followed by Late Triassic radiation in various invertebrate groups (Erwin 1993), and it therefore appears likely that the neoasteroids originated very early in the Triassic (Gale 1987). It is now necessary to review the record of Triassic asteroids to place minimum dates on the nodes of the cladogram (Text-fig. 20).

Asteroids from the Triassic of Germany; status of the Tricherasteropsida

The Triassic asteroid record includes six described species placed within four genera (Blake *et al.* 2000, 2006; Blake and Hagdorn 2003), of which three (*Migmaster* Blake, Bielert and Bielert, 2006, *Berckhemeraster* Blake and Hagdorn, 2003 and *Trichasteropsis*) have been placed in a family called the Trichasteropsidae Blake and Hagdorn, 2003. This family has been assigned to the order Trichasteropsida (Blake, 1987, which Blake (1987, fig. 10) identified as belonging to the Forcipulatacea and representing the monophyletic sister group to all other neoasteroids. The fact that the fossils are the earliest known Triassic asteroids forms part of the implicit evidence for the basal position of forcipulatids in the neoasteroid tree (Blake 1987; Blake and Hagdorn 2003). However, questions need to be asked about the monophyly of the trichasteropsids, the affinities of constituent genera and status of the Trichasteropsida.

The best-known trichasteropsid is *Trichasteropsis weissmanni* (von Münster, 1843), known from more than 30 specimens from the Muschelkalk of southern Germany. The arms are moderately produced, and a single marginal row is present, with intercalated intermarginals. The marginals increase in size to the arm tips, which are slightly swollen. The marginals are notched abactinally for passage of papulae. The abactinal skeleton of the disc has a reticulate, filigree form, with radial plates conjoined to interradial calcified septa by transverse struts. Distal radial ossicles are enlarged (Blake and Hagdorn 2003, fig. 1B, E, H). The actinal surface shows triangular interradial areas occupied by actinal ossicles in rows parallel with the marginals. The adambulacrals are broad and short and become broader distally along the length of the ambulacral groove, to occupy the entire breadth of the arm; they carry three to eight prominent spines arranged in a single transverse row (Blake and Hagdorn 2003, figs 1D, 2A–B). The oral ossicles are triangular; they bore five spines. A short adoral carina is present (Blake and Hagdorn 2003, fig. 1G). Details of orals, adambulacrals and ambulacrals are unknown, and pedicellariae have not been found.

The affinities of *Trichasteropsis* are controversial. Spencer and Wright (1966) included the genus in the Palaeozoic family Palasterinidae, Gregory 1899 on comparison of overall body form and the presence of interradial arcs without differentiated marginals. Hotchkiss and Clark (1976) included the genus in the extant family Poraniidae, because the growth gradients in the actinal ossicles of *Trichasteropsis weissmani* (MRP pattern of Blake and Hotchkiss 2004) compared well with those developed in that family. McKnight (1975) considered *Trichasteropsis* to be a paxillosid. Blake and Hagdorn (2003, fig. 7) included *Trichasteropsis* in a new family, the Trichasteropsidae, and concluded that it was a forcipulatid, nesting in a clade which included *Pisaster* Müller and Troschel, 1840 and *Zoroaster* Wyville Thomson, 1873. They provided the following reasons for the forcipulatid affinity of the genus:

– The tube foot row numbers are incipiently quadriserial;
– The presence of an adoral carina;
– The first adambulacral abuts the distal side of the oral;
– The oral ossicles are not robust or rectangular in ventral outline and are truncated proximally;
– The abactinal ossicles have a reticulate arrangement;
– The single marginal row compares well with the condition in Zoroasteridae.

However, few, if any, of these characters are exclusive to or even particularly characteristic of the Forcipulatida. Quadriserial tube foot arrangement occurs across a wide range of neoasteroid families (Korethrasteridae, Pterasteridae, etc.), and adoral carinae are developed in a diversity of nonforcipulatid asteroids such as the Jurassic *Plumaster ophiuroides* Wright, 1861 and some echinasterids. The shape of the orals and articulation between the oral and first adambulacral (Blake and Hagdorn 2003, fig. 1G) are quite dissimilar to the condition in extant forcipulatids in which the actinal face is very small and only carries few

spines; the orals in *Trichasteropsis weissmanni* have an elongated actinal surface with numerous spine bases. The reticulate arrangement of abactinals is closely similar between *Trichasteropsis* and extant *Porania*; compare the specimen of *Trichasteropsis weissmanni* figured by Blake and Hagdorn (2003, fig. 1B) with that of *Porania antarctica* illustrated by Hotchkiss and Clark (1976, pl. 1, upper figure). A single marginal row is present in various neoasteroids. In spite of the superb preservation of superficial spines on some specimens of *T. weissmanni*, not a single pedicellaria has ever been found. However, all forcipulatids possess forcipulate pedicellariae.

In conclusion, the affinities of *Trichasteropsis weissmanni* remain ambiguous; it shows characters present in, but not unique to, various extant neoasteroid families. It also has a number of unique characters of its own, such as the distally broadening adambulacrals. In the author's opinion, the species cannot currently be assigned to any suprafamilial group. Material of well-preserved dissociated ossicles should provide valuable evidence of the affinities of the species, in particular the numerous characters of orals, circumorals, adambulacrals and ambulacrals described for extant taxa in the present study.

Trichasteropsis bielertorum Blake and Hagdorn, 2003 (see also Blake *et al.* 2006) differs on important respects from the type species. The distal marginal ossicles are enlarged, and the arm tips are swollen (Blake and Hagdorn 2003, fig. 4) as in the Cretaceous stauranderasterid *Manfredaster* (see Villier *et al.* 2004c). The basic morphology is not dissimilar to that of *Trichasteropsis weissmanni*, with a single marginal row, distally broadening adambulacrals and actinals extending as far as the swollen distal arm. However, the abactinal construction is different in *Trichasteropsis bielertorum* to that in *Trichasteropsis weissmanni*, and the surface is made up of relatively large, irregularly sized polygonal ossicles rather than a reticulum of small, bar-shaped ossicles. *Trichasteropsis senfti* Blake and Hagdorn, 2003 (their fig. 3A–F) has elongated arms, narrow adambulacrals and small distal marginals.

Migmaster angularis Blake, Bielert and Bielert, 2006 is also from the Muschelkalk in Lower Saxony and is of Anisian age. The material comprises a larger individual of which the well-preserved actinal surface is exposed (Blake *et al.* 2006, fig. 4A–C; holotype, MHI 1809) and three smaller paratypes, two exposing the abactinal and one the actinal surfaces (MHI 1806, 1807 and 1808). The holotype has moderately produced arms and short, triangular interareas. The adambulacrals are very broad, with discrete lateral and actinal surfaces, which carry a small number of flattened, lanceolate spines. Large facets for interadambulacral muscles are present. A single marginal row can be seen, comprised of flattened, conical, distally directed ossicles that appear to become larger to the mid-arm. The interareas contain one or several larger, interradially positioned ossicles that do not contact the orals and probably are enlarged interradial marginals. The smaller paratypes are of a very different constructional type, being nearly pentagonal with very short arms and conspicuous rows of large supero- and inferomarginals. An enlarged single interradial inferomarginal is also present (Blake et al. 2006, figs D–E).

Blake et al. (2006) identified *Migmaster* as a neoasteroid belonging to the Trichasteropsidae, but with a mixture of characters of Palaeozoic asteroids and neoasteroids. They interpreted the interradial actinal plate as an axillary, or axillary like, ossicle. There are a number of problems with the published interpretation, the major one being the major constructional differences between the larger holotype and much smaller paratypes. Major ontogenetic changes can occur in asteroids (e.g. the oreasterid *Culcita novaeguinae* Müller and Troschel, 1842 which has a flat pentagonal disc when small, developing an inflated abactinal surface and subspherical form later in life (see Clark and Rowe 1971) but in the absence of any intermediate forms such interpretation is highly speculative in fossil material. It is best to treat the two taxa as separate and retain the name *Migmaster angularis* for the larger holotype.

The smaller paratype (MHI 1808) clearly has two marginal rows (Blake et al. 2006, fig. 3F), which precludes its inclusion in the Trichasteropsidae. The enlarged interradial actinal/marginal plate is not homologous with the axillary of Palaeozoic asteroids because it does not articulate with the orals (Text-fig. 14 herein). The adambulacrals of the holotype of *Migmaster angularis* are reminiscent of those seen in Permian asteroids from Australia, which have a similar angulation and carry a comparable number and type of large spines (Kesling 1969, pls 2–3).

Berckhemeraster charistikos Blake and Hagdorn, 2003 is poorly preserved, but appears to have two marginal rows (the 'actinal' and marginal rows of Blake and Hagdorn 2003) and short, broad adambulacrals that become narrower distally (Blake and Hagdorn 2003, fig. 5A–B). The abactinal ossicles are polygonal and form a tessellation. In the author's opinion, this genus should not be referred to the Trichasteropsidae.

In conclusion, the Muschelkalk asteroids are true neoasteroids because they have an internal odontophore not in contact with the marginals, possess actinal ossicles and show ambulacral construction of neoasteroid type (Blake and Hagdorn 2003, fig. 3G–H). The presence of spaces for intermarginal papulae and of intermarginals (see characters 25, 35) indicate that the Muschelkalk asteroids belong to the Surculifera. They include two separate morphological groupings, quite distinct from other neoasteroids, which are justifiably characterized as discrete families. First, the Trichasteropsidae Blake and Hagdorn, 2003 can be defined as a group with moderately produced arms; a single row of marginals that enlarge towards the arm tip; short, broad adambulacrals that have broadly curved lateral and actinal external surfaces and that broaden by two to three times towards the arm tip; they carry a single transverse row of large spine bases. Second, a very different morphological group is here called the Migmasteridae fam. nov. (see below, section 'Systematic Palaeontology'). These possess short, rapidly tapering arms; blocky paired supero- and inferomarginals; a large, single interradial inferomarginal which is inset from the ambitus; oval transversely broad adambulacrals that carry four or five transversely arranged, blade-like spines adradially; abradially, they each bear a single large, blade like spine on the distal margin.

There is little support for the idea that either the Muschelkalk asteroids form a sister group to all other neoasteroids as indicated by Blake (1987), or *Trichasteropsis weissmanni* is more closely related to the extant forciculatids *Pisaster* and *Zoroaster* than to *Pedicellaster* M. Sars, 1861 (Blake and Hagdorn 2003). The diagnostic characters of the Trichasteropsida given by Blake and Hagdorn (2003) cannot be used to characterize asteroids of the Muschelkalk. The Trichasteropsidae and Migmasteridae are here treated as Neoasteroidea *incertae sedis*. The discovery of well-preserved isolated ossicles of the ambulacral groove and mouth frame should enable the relationships of the Muschelkalk asteroids to be determined more precisely.

The single specimen of *Noriaster barberoi* Blake, Tintori and Hagdorn, 2000 from the Upper Triassic (Norian) of northern Italy possesses large, flattened inferomarginals and broad actinal interareas made up of flat, spine-bearing ossicles. The adambulacrals are short and rectangular, and each carried two spines. The abactinal surface includes enlarged superomarginals and radial ossicles (Blake et al. 2000, figs 1–3). *Noriaster barberoi* was assigned by Blake et al. (2000) to the Poraniidae, largely on the basis of similarities of ossicle arrangement of the actinal surface with that family. However, the presence of large superomarginals and abactinal ossicles (their fig. 3b, d) is quite unlike that seen in any living member of the family, and *Noriaster* is here also considered to be a neoasteroid of uncertain affinity.

Asteroid ossicles from the Carnian of St Cassian, Cortina d'Ampezzo, Italy

Dissociated but extremely well-preserved asteroid ossicles from the lower Carnian (Upper Triassic) St Cassian Formation at Cortina d'Ampezzo, northern Italy, were figured by Zardini (1973, pl. 20). These include an oral, an adambulacral, an ambulacral, marginals and abactinal ossicles, which Gale (1987) referred to the Asteriidae and Goniasteridae, but which Blake (1987) and Blake and Hagdorn (2003) more cautiously considered to be neoasteroid but of uncertain affinities. Recent surface picking and bulk sampling has yielded copious new material from the locality called Milieres, west of Cortina (Bizzarini *et al.* 1989). The ossicles are found in clays that contain an abundant fauna of calcitic benthos (corals, molluscs, echinoids, sponges), thought to have been washed from a reef platform into deeper, muddy water.

The commonest adambulacral type in the St Cassian material (type A; Pl. 16, figs 1–3, 12; Text-fig. 26) is rectangular in proximal/distal aspect and shorter than broad. The abactinal face has *ada1a*, *ada1b*, *ada2* and *ada3* set symmetrically on either side of the *padam* and *dadam* muscle attachment sites. The actinal face carries dense spine bases set in 4–5 longitudinal rows and the abradial face has a vertically oriented row of 3–5 bosses close to the distal margin of the ossicle. The abactinal arrangement of articulation surfaces and muscle sites is very similar to the arrangement seen 'valvatids' such as the living ophidiasterid *Nardoa variolata* (Text-fig. 26A, B), but the very high density of spines on the actinal face (each ossicle must have borne 20–30 subadambulacral and adambulacral spines) and the vertical row of bosses on the proximal abradial face are not known in any living species. A common ambulacral type (A) in the St Cassian material (Pl. 16, fig. 6; Text-fig. 26D) shows all the characters of a neoasteroid ambulacral ossicle. There is moderate elongation and imbrication of the proximal part of the ambulacral head, and the dentition is restricted to two to three central, vertically oriented, tall bars. The *abtam* attachment site is short and deep. The actinal surface of the ambulacral base shows short, near-symmetrical wings for *padam* and *dadam* attachment and clearly defined articular structures with the adambulacral (*ada1a*, *ada1b*, *ada2*, *ada3*), symmetrically placed on proximal and distal sides of the transverse actinal ridge. The close similarity in size, shape and arrangement of structures on the actinal face of the ambulacral base and the abactinal surface of the adambulacral type A described above afford evidence that they belong to the same species. The ambulacral morphology is perhaps closest to that seen in extant ophidiasterids (*Nardoa variolata*; Text-fig. 26A), but the

short, tall dentition and deep *abtam* site are not found in modern species.

Oral ossicle type A is the commonest in the material (Pl. 16, figs 7–8; Text-fig. 26F), and similarities of spine arrangement between the actinal face of orals and type A adambulacrals provide some evidence that they originated in the same species. The oral ossicles have a triangular, elongated proximal blade (*pb*) that possesses a flat radial articulation with the oral from the adjacent pair. The apophyse is low, with a shallow, rather horizontal groove for the ring vessel. Interoral articulations are present in the interradial face of the proximal bar. Both proximal (*poda*) and distal (*doda*) odontophore articulation facets are seen on the distal interradial surface, and an odontophore capsule for insertion of the odontophore-oral muscle is present (*odc*). The radial (outer) face of the oral has a large actinal face that is slightly convex and carries numerous spine bases in 10 poorly defined, oblique rows. Oral type A is similar to orals of various 'valvatids' in shape and construction, including *Archaster lorioli* (Text-fig. 26C), but the apophyse is not so strongly inclined proximally, and the spines of the actinal face are much denser.

A single circumoral ossicle was collected (Pl. 16, fig. 9), which is possibly associated with oral type A. This had parallel-sided proximal and distal processes; the proximal process is short and the distal one elongated. The morphology is closest to *Nardoa* Gray, 1840, *Mediaster* Stimpson, 1857 and *Archaster*.

Many close similarities exist between ossicles of the extant 'valvatids', including the ophidiasterid *Nardoa variolata*, and the adambulacral, ambulacral and oral ossicle types A from the Carnian clays of Cortina d'Ampezzo. However, there are small but significant differences between the two, in oral and adambulacral spine arrangement and density and in the construction of the ambulacral head. It is not presently known whether these are paralleled in any extant ophidiasterid or represent a unique morphological type.

Oral type B (Pl. 16, figs 4–5) has a very tall vertical apophyse, which although chipped is clearly seen to carry a tall vertical groove for the ring vessel on the interradial

---

EXPLANATION OF PLATE 16

Asteroid ossicles from the Carnian (Middle Triassic) St Cassian Formation of Milieres, Cortina d'Ampezzo, Italy.

Figs 1–3, 6–9, 12. Undescribed taxon, close to Ophidiasteridae. Adambulacral ossicles in abactinal (Figs 1–2), distal (Fig. 3; NHM EE 13580) and proximal views (Fig. 12) (Fig. 1, NHM EE 13578; Figs 2, 12, NHM EE 13579). Ambulacral ossicle (Fig. 6; NHM EE 13582) in actinal aspect. Circumoral ossicle (Fig. 9; NHM EE 13583) in abactinal view. Oral ossicle (Figs 7–8; original of Zardini, 1971, pl. 20, fig. 14a, b; Zardini Collection, Cortina d'Ampezzo) in radial and interradial aspects, respectively.

Figs 4–5. Spinulosid oral ossicle (NHM EE 13581), in radial and interradial aspects, respectively. Compare with specimens in Text-figure 18C, E.

Figs 10–11. Undescribed taxon, possible stauranderasterid, oral ossicle (NHM EE 13584) in radial and interradial views, respectively.

Scale bar represents 2 mm.

PLATE 16

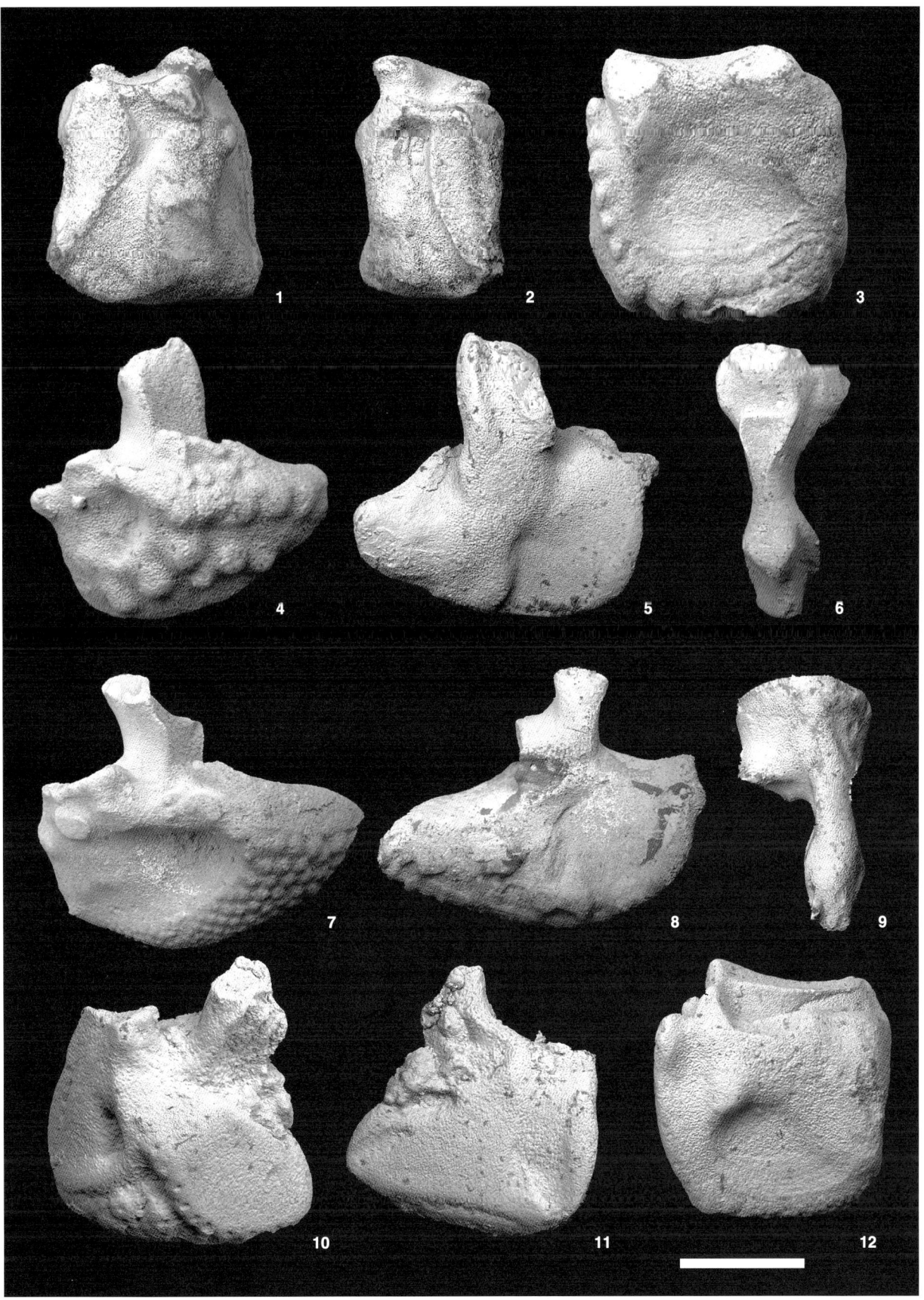

GALE, Triassic asteroid ossicles

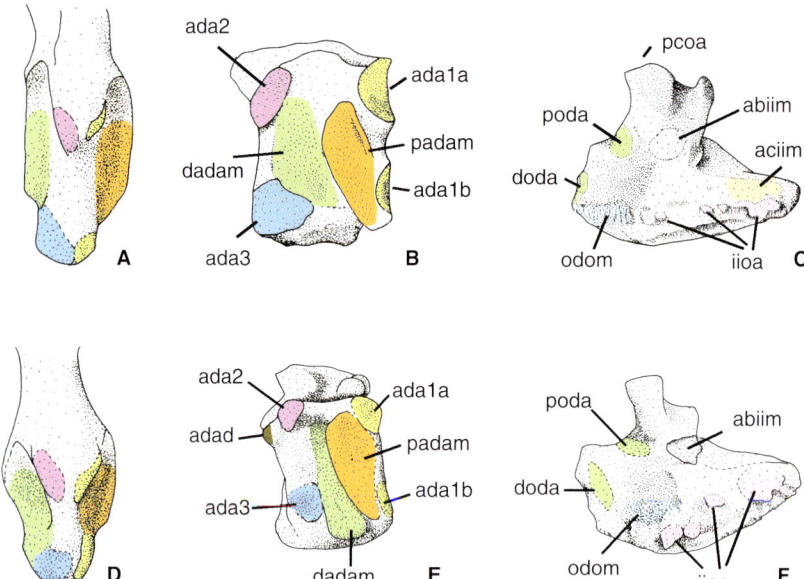

TEXT-FIG. 26. Asteroid ossicles from the St Cassian Formation of Milieres, Cortina d'Ampezzo, Italy (lower Carnian); comparison with the extant ophidiasterid *Nardoa variolata* and the archasterid *Archaster lorioli*. A, B, *N. variolata*, ambulacral base in actinal aspect (A) and adambulacral ossicle in abactinal aspect (B). C, Oral ossicle of *Archaster lorioli* in interradial aspect. D–F, undescribed taxon, lower Carnian, St Cassian Formation. D, actinal view of base of ambulacral. E, abactinal aspect of adambulacral. F, interradial view of oral ossicle. Ossicles coloured to indicate homologous structures (compare with Text-figs 10, 11). A, B, D, E, ×20; C, F, ×6.

face. The proximal blade is short, and the posterior part of the inner face is concave and modified for odontophore articulation and insertion of the odontophore-oral muscle. The radial face (Pl. 16, fig. 4) shows a proximal rim that carried a row of oral spines and a separate, gently convex actinal surface that shows bosses for the attachment of 5–6 suboral spines. The oral type B falls closest to *Asterina gibbosa* of the extant species figured herein (Pl. 13, figs 4–5), with which it shares the tall apophyse carrying a long vertical ring vessel groove and the construction and spination of the radial face, with well-separated oral and suboral regions and spine attachments. No other ossicle types likely to be associated with this oral are currently known. It is referred to the Spinulosida *incertae sedis*, with possible affinities to the Asterinidae. This will require investigation into the oral morphology of a wider range of taxa than included in this study.

Dating the neoasteroid radiation

The St Cassian material is important because it provides the only evidence of detailed asteroid ossicle morphology known from Triassic strata and therefore permits minimum dates to be assigned to nodes on the neoasteroid cladogram, irrespective of the precise taxonomic assignation of the material (Text-fig. 26). Thus, the basal neoasteroid split into Paxillosida and Surculifera (Text-fig. 21, node 48) had taken place by the Anisian because of

evidence from the Muschelkalk taxa (see above). The division of the Surculifera into Spinulosida and Tripedicellaria (Text-fig. 21, node 47) must have occurred prior to or during the early Carnian, because both groups are present in the St Cassian material. Finally, it is likely that a large part of the diversification of the basal Tripedicellaria (Text-fig. 21, nodes 39–41) had occurred by the Carnian.

The St Cassian material is also remarkable because it provides evidence that an asteroid closely related to the Ophidiasteridae was living on Triassic Tethyan reefs. Ophidiasterids are most abundant and diverse on modern-day tropical reefs (Downey, *in* Clark and Downey 1992), and the densely spinose, tightly closed nature of the groove and mouth regions and the close granule cover of the dorsal surface provides armour to protect the vulnerable papulae, tube feet and peristomial tissues in an environment where extensive grazing takes place (Blake 1983). The St Cassian material thus provides evidence that at least part of a reef ecosystem of modern aspect was present by the Carnian, only about 20 million years after the end of the Permian.

The almost simultaneous appearance of diverse asteroid morphologies of very modern aspect, many assignable to modern families, in the Lower to Middle Jurassic (Hettangian–Bajocian) is a striking feature of the asteroid fossil record (Text-fig. 27). It has been assumed that this sudden appearance of taxonomically diverse asteroids represents a very rapid evolutionary radiation of latest Triassic or earliest Jurassic age, but evidence that diverse

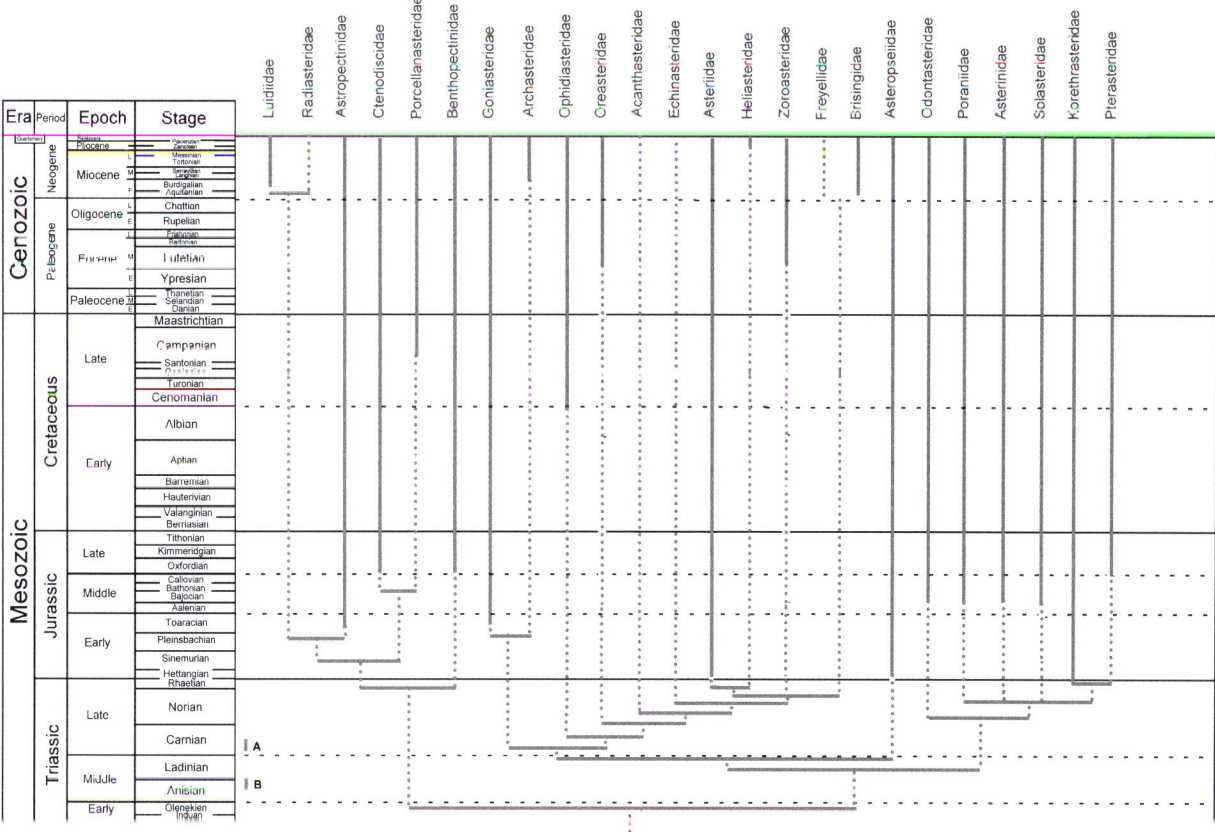

TEXT-FIG. 27. Stratigraphical distribution of neoasteroid families in relation to phylogeny given in Text-figures 20, 21. The stratigraphical position of the Triassic faunas from the St Cassian Formation of Italy (A) and the Muschelkalk of Germany (B) are shown.

neoasteroid faunas were already present in the Carnian (earliest Late Triassic) casts doubts on this.

Permian and Triassic marine sediments are underrepresented in the stratigraphical record, probably because assembly of the supercontinent Pangea significantly reduced the global shelf-sea area (Smith and McGowan 2007). In contrast, the Early Jurassic was marked by a major and long-lived transgression, related to the early opening of the Atlantic Ocean and breakup of Pangea (Hesselbo 2000), and Hettangian marine rocks overlie Rhaetian marginal or nonmarine facies across much of Europe and locally onlap onto basement. Thus, the extent and outcrop areas of Lower Jurassic marine sediments are far greater than for those of Triassic age, creating an artefact of marine sediment distribution, and hence of fossil preservation (Smith 2001; Smith and McGowan 2007). It is probable that this is the real cause of the appearance of diverse neoasteroids in the earliest Jurassic. Furthermore, the preservational window provided by the St Cassian ossicles provides good evidence that much of the radiation of the neoasteroids had taken place by the Late Triassic.

There are a significant number of long ghost lineages in the neoasteroid record (Text-fig. 27), of which the most remarkable are: (1) the absence of Paxillosida until the

Early Jurassic (Toarcian) and (2) the absence of Luidiidae until the Miocene, because the phylogeny implies a Jurassic origin for the common ancestor of *Radiaster* and *Luidia*. Similarly, the absence of any convincing fossil Echinasteridae is surprising, because the group should have originated in the Triassic. The Archasteridae also is absent until the Miocene, and this family also has an inferred Jurassic origin. The Mesozoic taxa referred to the Forcipulatida need revision to clarify their correct affinities (see above). An apparent gap in the fossil record of the Asterinidae between the Bajocian *Mesotremaster* (Fell 1972) and the abundant and diverse representation of the family at the present day is also remarkable. However, as time progresses, more Jurassic records of families previously only known from younger sediments are being confirmed.

REAFFIRMATION OF THE BASAL POSITION OF THE PAXILLOSIDA

The phylogenetic position of the Paxillosida within the Asteroidea has been the basis of a longstanding (nearly one hundred years) controversy, which commenced with a disagreement between MacBride (1921) and Mortensen

(1921) over the significance of asteroid larval stages. Mortensen interpreted the presence of a brachiolaria larva as an advanced condition within the asteroids, and its absence in paxillosids was therefore evidence for their primitive position. MacBride argued that the presence of a brachiolaria was the primitive condition within the Asteroidea and was secondarily lost by paxillosids, because the adhesive brachiolarian arms were of no use for attachment to the soft substrates inhabited by paxillosids.

The basal position of the paxillosids within the entire asteroid class was supported by Fell (1963), who interpreted the extant *Platasterias latiradiata* (Gray, 1871) as a 'living fossil' somasteroid surviving from the Ordovician, rather than a luidiid asteroid, its conventional taxonomic position. Fell also placed the remaining Luidiidae in the Palaeozoic order Platyasterida, and his classification was developed further by Spencer and Wright (1966, fig. 38), who integrated Perrier's classification of extant asteroids with Spencer's Palaeozoic groupings. Thus, the four extant orders Paxillosida, Spinulosida, Valvatida and Forcipulatida were traced back into the Ordovician, with Platyasterida (including the Luidiidae) and Paxillosida as basal to all asteroids. Fell's (1963) interpretation of *Platasterias latiradiata* as a surviving somasteroid was subsequently challenged by Madsen (1966) and Blake (1972, 1982), who demonstrated from skeletal morphology that the species was a true *Luidia* and therefore a member of the Paxillosida.

In 1987, Blake and Gale independently published new phylogenies of the asteroids in which the post-Palaeozoic taxa were identified as a monophyletic group, called the Neoasteroidea by the latter author (Gale 1987). However, the phylogenies differed fundamentally in that Gale placed the Paxillosida as basal to the neoasteroids, whereas Blake identified the Forcipulatida as the most plesiomorphic group and interpreted the Paxillosida as secondarily simplified valvatids. Thus, the MacBride/Mortensen controversy was reborn, and Blake (1988a, b) developed the argument originated by McBride that paxillosids are highly specialized for living on soft substrata and have secondarily lost numerous features. Heddle (1995) provided criticism of the argument and reiterated the arguments for the basal position of the paxillosids.

The development of molecular phylogeny enabled independent testing of the relationships of the Paxillosida with other asteroids. Lafay *et al.* (1995) investigated the 28S ribosomal RNA sequences of nine asteroid species representing the major groups and combined their data with morphological data to generate trees. They concluded that the Paxillosida were basal, but not monophlyetic. Wada *et al.* (1996) developed a phylogeny based on mitochondrial rDNA and concluded that the Paxillosida were also paraphyletic and basal. A subsequent study by Knott and Wray (2000) proved inconclusive. A study of the complete mitochondrial DNA nucleotide sequences of five asteroids (Matsubara *et al.* 2005) concluded that paxillosids were monophyletic, but not basal neoasteroids and that their characters were secondarily derived. Molecular evidence thus provides ambiguous evidence for both paxillosid relationships and the monophyly of the group.

The present study has investigated in detail the morphological evidence which is relevant to the phylogenetic position of the paxillosids and concluded that they are indeed basal neoasteroids. The new evidence is summarized here, together with a review of evidence from soft-part and developmental studies.

1. In paxillosids, the adambulacrals of a single row articulate by means of the paired surfaces *adad* and *ada3*; the *ada3* articulation is exclusively between adambulacrals (Text-fig. 7; see also Blake 1973). An identical means of articulation is seen in Late Palaeozoic asteroids (e.g. *Calliasterella mira*; Text-fig. 7B, C). However, in all nonpaxillosid neoasteroids, the *ada3* surface shares an articulation with the *ada3* surface of the proximal ambulacral (Text-fig. 8), a fundamental difference in the nature of the ambulacral–adambulacral structure.

2. Paxillosids possess a single distal adambulacral–ambulacral articulation surface (*ada1*; see also Blake 1973), as do Late Palaeozoic asteroids such as *Calliasterella mira* (Text-fig. 7B, C). All other neoasteroids (with the exception of *Odontaster validus*) have discrete ab- and adradial facets (*ada1a*, *ada1b*) on the adambulacral and ambulacrals.

3. The ambulacral heads of paxillosids are upright and nonimbricating, and transversely nearly symmetrical (Pl. 5; Text-fig. 28A), the same condition as observed in all Late Palaeozoic asteroids (e.g. Text-fig. 28B). The ambulacral heads of all other neoasteroids (Surculifera) imbricate proximally to a varying degree and are transversely strongly asymmetrical (Pl. 5; Text-fig. 8C).

4. The distal circumoral process of all paxillosids is angled at 45–75 degrees to the body of the circumoral ossicle, whereas the process is parallel with the body of the ossicle in other neoasteroids (Text-figs 19, 29). The distal circumoral process has a similar angle to that seen in paxillosids in the three Carboniferous asteroids for which this ossicle is known (e.g. *Delicaster enigmaticus* (Kesling, 1967), Blake and Elliot, 2003, pl. 3, fig. 4; Text-fig. 29F herein; undescribed Carboniferous asteroid from Ireland, Text-fig. 29E).

5. The presence of an external, spine-bearing face on the odontophore in many paxillosids (all goniopectinids, porcellanasterids and luidiids; some astropectinids; Text-fig. 30A, B, E) is a plesiomorphic character,

TEXT-FIG. 28. Morphology of ambulacral heads of Palaeozoic and Recent asteroids. A, benthopectinid *Pontaster tenuispinus*. B, undescribed genus, (original of *Palaeaster struchburii* Etheridge, 1898, pl 13, fig. 1). C, goniasterid *Ceramaster granularis*. Note that the ambulacral heads of the extant paxillosid (A) and the Palaeozoic asteroid (B) are nonimbricating and transversely symmetrical, with ambulacral angles approaching 90 degrees (compare with Text-fig. 5), whereas those of other neoasteroids (Surculifera) imbricate to a variable degree, are transversely asymmetrical and have lower ambulacral angles (C). All ×5.

which is also developed in all Late Palaeozoic asteroids. Paxillosids effectively represent a morphological transition from the Palaeozoic condition in which a large external face is present on the odontophore (Text-fig. 30D), to the surculiferan condition in which the ossicle is small and entirely internal (Text-fig. 30C). A similarly small external face on the odontophore is found on the stem group Permian species *Permaster grandis* (Text-fig. 30F).

6. Paxillosids possess only elementary pedicellariae; these comprise two to four simple valves conjoined by adductors and are attached to the body surface by abductor muscles (Text-fig. 3). Other neoasteroids possess only alveolar or complex pedicellariae, never elementary ones. Elementary pedicellariae are found in Late Palaeozoic species including the Middle Devonian *Arkonaster topotorum* Kesling, 1982 and the Silurian *Bdellacoma vermiformis* Salter, 1857 (see Sutton *et al.* 2005). The smaller three-valved type of pedicellariae found in *A. topotorum* (see Kesling 1982, pl. 6) are extremely similar to those described for extant

species of *Luidia* (see Clark and Downey 1992). Isolated valves of distinctive asteroid elementary pedicellariae are widespread in Silurian to Carboniferous sediments and have been described under the name *Bursulella* Jones, 1887 (Boczarowski 2001; Sutton *et al.* 2005). No evidence of alveolar or complex pedicellariae is known from the Palaeozoic.

7. The marginal position of the madreporite in paxillosids is remarkably consistent in extant forms and intermediate in position between the more central position of Surculifera and the more marginal position observed in all Late Palaeozoic asteroids. The ratio of the distance between the inner margin of the superomarginal and the distance between the centrale and madreporite (PM) was measured in approximately 100 neoasteroid and Late Palaeozoic species (Text-fig. 31). Species that lack clear marginals were not included. Paxillosids plot with a concentration of values in the range 1.5–3.0; higher values (3.0–7.5) are mostly for goniopectinids and porcellanasterids which have very marginal, often large, madreporites

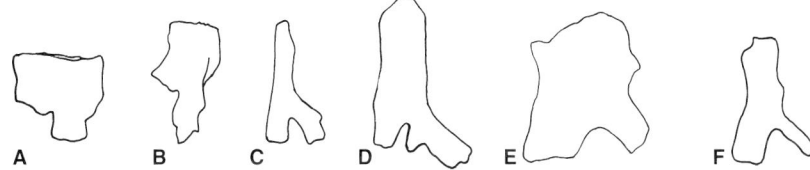

TEXT-FIG. 29. Comparative morphology of circumorals in oblique abactinal aspect. A, B, surculiferan circumorals. A, *Asterias rubens*. B, *Mediaster aequalis*. C, D, paxillosid circumorals. C, *Styracaster chuni*. D, *Radiaster tizardi*. E, F, Late Palaeozoic (Carboniferous) species. E, undescribed form from the Asbian of Gleniff, Sligo, Ireland. F, *Delicaster enigmaticus* (Kesling, 1967) (after Blake and Elliot 2003, pl. 3, fig. 4), Mississippian of Kentucky. In Late Palaeozoic forms and paxillosids, the distal circumoral bar is angled to the body of the ossicle. In surculiferans, the distal bar is vertical and the proximal bar very short, ×10.

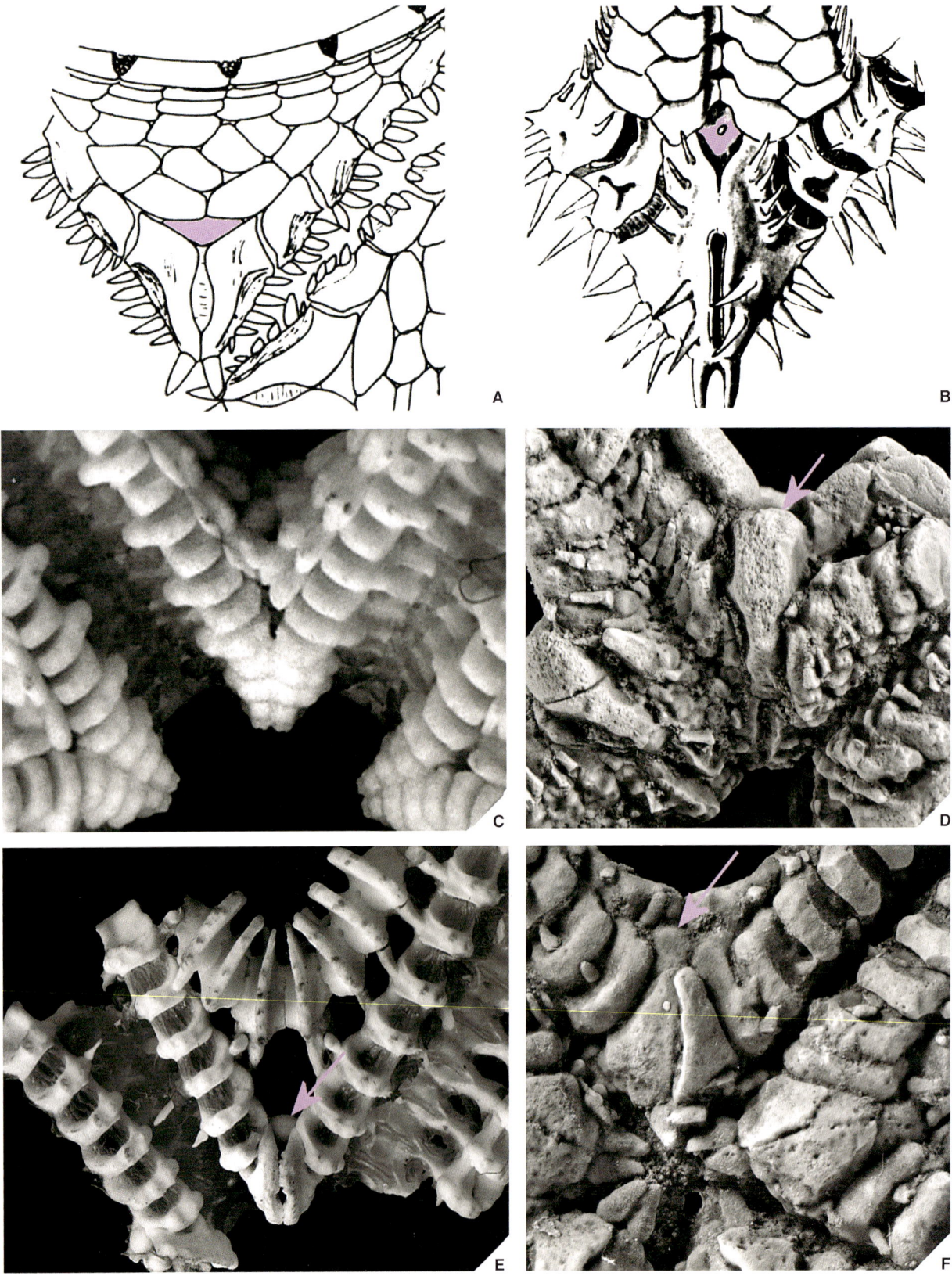

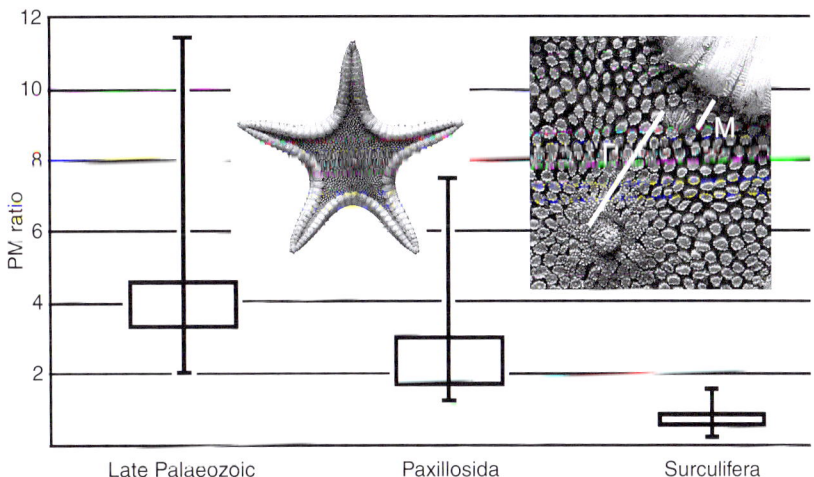

TEXT-FIG. 31. Plot of the relative position of the madreporite to the centre of the disc and the inner margin of the interradial superomarginal (PM ratio). There is a clear division between paxillosid species in which the madreporite is more marginal in position (PM >1), and surculiferan species in which it is more central (PM <1), which correlates with the early postmetamorphosis development of the abactinal skeleton following either the EMM or LMM (see Text-fig. 32), respectively. Interestingly, the Late Palaeozoic species had even more marginal madreporites than paxillosids and may well have had a similar mode of early development.

(Madsen 1961). Other neoasteroids (Surculifera) have equally consistent low PM values clustering between 0.7 and 0.9, with a very limited total range (0.3–1.6). Only a single species of surculiferan plotted in the paxillosid field (the ophidiasterid *Fromia* sp., 1.6). There is thus a significant and consistent difference in the position of the madreporite in neoasteroids between paxillosids and surculiferans. Late Palaeozoic species (Devonian, Carboniferous and Permian) clustered between 3.2 and 4.7, with a few individuals giving very high values (10, 11.5).

The position of the madreporite is significant because it appears to be determined by the nature of early post-metamorphosis development of the abactinal skeleton. Oguro *et al.* (1976) and Chia *et al.* (1993) demonstrated that there is a consistent difference in the developmental pattern displayed by paxillosids when compared to that seen in all other asteroids. In *Luidia* spp. (Luidiidae), *Astropecten* spp., *Ctenopleura fisheri* Hayashi, 1957 (Astropectinidae) and *Ctenodiscus australis* Lütken, 1871 (Goniopectinidae; see Lieberkind 1926), the madreporite appears at the earliest stage of ossicle formation immediately after metamorphosis, forming a ring with the five terminal ossicles on the margin of the abactinal surface

(Text-fig. 32E, F; here called the Early Madreporic Mode, EMM). The madreporite remains in a marginal position during the formation of the primary circlet of five primary interradials, five primary radials and centrale (Text-fig. 32F, G). In contrast, in all other asteroids studied (e.g. *Asterina* spp., *Asterias* Linnaeus, 1758, *Leptasterias* Verrill, 1866, *Certonardoa* H. L. Clark, 1921, *Pentaceraster* Döderlein, 1916; see Oguro *et al.* 1976), the madreporite appears later in development (Late Madreporic Mode, LMM), after formation of the centrale, five primary radials, five primary interradials and numerous other secondary abactinals and marginals (Text-fig. 32A–D). For example, in *Asterina minor* Hayashi, 1974 (Komatsu *et al.* 1979), the madreporite only appears after numerous abactinal ossicles have formed, with R of about 1.5 mm (Text-fig. 32D). In *Leptasterias ochtoensis similispinis* (H. L. Clark, 1908), the madreporite only appears as a rudiment 2 years after metamorphosis (Kano *et al.* 1974). The two conditions correlate precisely with nonbrachiolarian (EMM) and brachiolarian development (LMM), so there is a probably a fundamental difference in the way the postmetamorphosis skeleton develops from the bipinnaria/barrel-shaped larva and the brachiolaria.

TEXT-FIG. 30. Development of the odontophore/axillary in Late Palaeozoic (Permian) asteroids and neoasteroids. A, porcellanasterid *Styracaster horridus* Sladen 1883, after Madsen (1961, fig. 17b). B, goniopectinid *Ctenodiscus orientalis* Fisher, 1919 (pl. 7, fig. 1f). C, *Zoroaster fulgens* (Table 1). D, *Monaster carnarvonensis* Kesling 1969; UWA 26992d, original of Kesling (1969, pl. fig. 8). E, *Luidia* sp. F, *Permaster grandis* Kesling 1969, UWA 26992e, original of Kesling (1969, pl. 1, figs 9–11; pl 2, figs 1–6). The large odontophore/axillary in *Monaster carnarvonensis* articulates with the inferomarginals laterally and the orals proximally, as does the reduced ossicle in *Permaster grandis* which has a small external face. Small external faces on the odontophore exist in some paxillosids, between the actinals and orals, notably porcellanasterids (A), goniopectinids (B) and luidiids (E). Odontophore/axillary pink. All ×5.

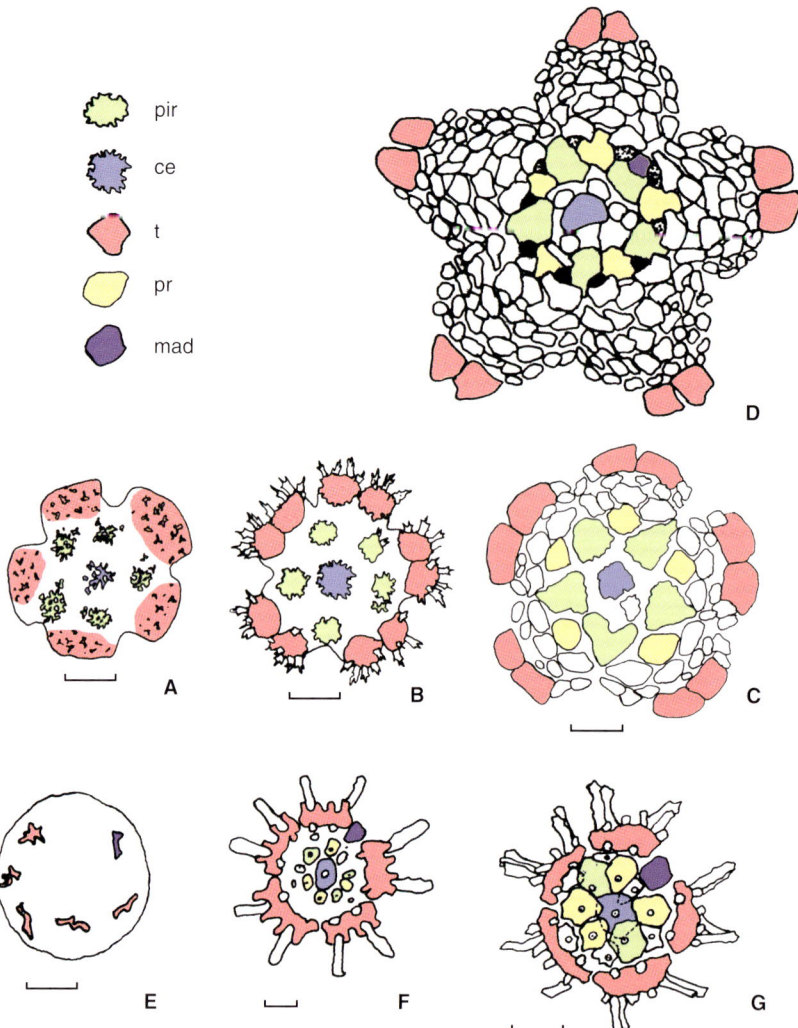

TEXT-FIG. 32. Postmetamorphosis development of the abactinal skeleton in neoasteroids. A–D, *Asterina minor*, after Komatsu *et al.* (1979, figs 7–11). E–G, *Astropecten latespinosus* Meissner, 1892, after Komatsu (1975, figs 28–34). In the Early Madreporite Mode (EMM, see E–G), the madreporite appears at the earliest phase of skeletogenesis, together with the five terminals (E) and remains in a marginal position. In the Late Madreporite Mode (LMM), terminals, primary radials and centrale appear together (A, B), then primary radials, radials and marginals (C). The madreporite appears adjacent to the *pir* of the II interradius at a later stage (D). Of the investigated species, astropectinids, goniopectinids and luidiids all display the EMM; all other neoasteroids show the LMM. Scale bar in A and B represents 100 μm; in C 150 μm; E, F 100 μm; G 75 μm and D 500 μm.

Whenever it develops, the madreporite is consistently positioned immediately distal to, or even incorporated within, the primary interradial plate of the interradius II–III (e.g. Komatsu *et al.* 1979). There is a consistent relationship between asteroids in which the madreporite is more marginal in position (>1.3 MP) and those with Early Madreporic Mode of development (Text-fig. 32E, F), and conversely, those with more central madreporites and Late Madreporic Mode development. The precise control is unknown, because no one has yet followed the full developmental history of the asteroid abactinal skeleton. However, the explanation probably lies in the way in which abactinals are introduced during formation of the late juvenile and adult skeleton. In the EMM, many more new abactinal ossicles must be introduced within the primary circlet, effectively pushing the primary interradial II–IV and the madreporite into a more marginal position. Conversely, in the LMM, most new abactinals must form between the primary circlet and the margin, making the madreporite more central in position (e.g. Text-fig. 32A–D). The close similarity in the position of the madreporite between Late Palaeozoic species and paxillosids supports the basal position of the latter group and makes it very likely that the EMM of abactinal development was plesiomorphically shared by Late Palaeozoic asteroids and paxillosids.

8. The absence of suckered tube feet in paxillosids has long been considered a primitive character (e.g. Fell 1963; Gale 1987). However, Blake (1988*a*, *b*) used the observations of Oguro *et al.* (1976) that certain juvenile astropectinids possess suckers to argue that paxillosids have secondarily lost these structures as an adaptation to dwelling on soft substrata. In a more recent study of asteroid tube feet, Vickery and McClintock (2000) proposed a new classification combining the shape of the tip with the presence or absence of suckers. Representatives of five paxillosid families (Luidiidae, Astropectinidae, Goniopectinidae Verrill, 1889, Porcellanasteridae, Benthopectinidae) were investigated, and no suckers were found on either juvenile or adult individuals. Tube feet in the first four families were classified as 'pointed non-suckered' (PNS), and 'semipointed non-suckered' (SPNS) in benthopectinids. Valvatid families (Odontasteridae, Ganeriidae, Sladen, 1889, Asterinidae, Poraniidae) possess flat nonsuckered tube feet (FNS) and suckers are found only on solasterids, korethrasterids, pterasterids, echinasterids, zoroasterids, labidiasterids and asteriids. The situation is a complex one, but Vickery and McClintock (2000) broadly supported the plesiomorphic position of paxillosids. Tube feet are preserved on a few Palaeozoic asteroids and, if forcipulatids really are basal neoasteroids, their ancestors should also have had suckered tube feet. Gale (1987) described and figured putative pointed tube feet on the Silurian asteroid *Siluraster caractaci* (Gregory 1899), and Sutton *et al.* (2005) illustrated definitively pointed nonsuckered tube feet on the Silurian asteroid *Bdellacoma vermiformis* Salter, 1857.

9. Paxillosids possess the most simple digestive systems of any asteroids, but as a group display considerable variation in the development of digestive structures (Jangoux 1982). Luidiids and porcellanasterids all lack an anus, a rectum and pyloric caecae, as do most goniopectinids and astropectinids including *Astropecten* itself. However, some (mostly deeper-water) astropectinids such as *Thrissacanthias* Fisher, 1910, *Dipsacaster*, *Dytaster* Sladen, 1889, *Psilaster* Sladen, 1885, *Persephonaster* Wood-Mason and Alcock, 1891, *Plutonaster* Sladen, 1889, plus the goniopectinid *Prionaster analogus* Fisher, 1913, all possess an anus, rectum and rectal caecae. Paxillosids are incapable of extraoral feeding by complete stomach eversion, and the stomach retractor system is very poorly developed, consisting of a few securing strands (Jangoux 1982). Blake (1988*a*, *b*) argued that independent evolutionary loss of structures is easier than parallel evolution of an anus, rectum

and rectal caecae in separate lineages. However, Jangoux (*in* Blake 1988*a*, *b*) was of the opinion that the loss of rectal caecae, which perform a unique excretory function, was difficult to explain. A preserved digestive system in the Silurian asteroid *Bdellacoma* (Sutton *et al.* 2005) shows a simple construction, rather like *Luidia,* with no anus, caecum or pyloric caecae.

10. The absence of a brachiolaria larva in paxillosids has been known for a long time, and disagreement over the significance of this triggered the original debate over the phylogenetic position of the paxillosids (MacBride 1921; Mortensen 1921). A very useful review of developmental modes was provided by Wada *et al.* (1996), who followed Oguro (1989) in recognizing four developmental modes in asteroids, comprising: (1) direct mode (brachiolaria, no bipinnaria); (2) indirect mode (successively, bipinnaria and brachiolaria); (3) non-brachiolarian mode (bipinnaria only); and (4) mode with barrel-shaped larvae, no bipinnaria or brachiolaria. Paxillosids show only modes 3 and 4 and never have a brachiolaria; in contrast, all other asteroids possess a brachiolaria larva. The brachiolaria possesses three processes called brachiolarian arms surrounding a central sucker, which together are used to settle on the substrate during metamorphosis. Blake (1988*a*, *b*) reiterated the argument of MacBride that the brachiolarian arms would be of little use in settling on the soft substrata inhabited by paxillosans, and therefore the brachiolarian larval stage was lost in the group. The contention that paxillosids are secondarily derived leads McEdward and Janies (1993) to propose that the presence of a brachiolaria is the primitive condition in asteroids thus resurrecting MacBride's (1921) original argument. In the complete absence of any fossil evidence, it is effectively impossible to test the status of larval development without reference to molecular phylogeny (Wada *et al.* 1996).

If paxillosids are in fact a derived group, which have secondarily lost numerous characters, they have independently evolved ambulacral groove articulations and morphologies, circumoral ossicle morphologies, a lateral madreporite and an exposed odontophore homoplastically with Late Palaeozoic asteroids. They have secondarily lost the more efficient types of pedicellaria, suckered tube feet and most of the digestive system, including the anus and rectal caecae, which are important in asteroid excretion. Additionally, they have lost the brachiolaria larva. Conversely, if forcipulatids are basal neoasteroids, they have evolved the most intricate skeletal construction, complex pedicellariae and digestive systems at the earliest stage of neoasteroid development.

HOMOPLASY AND PARAPHYLY IN ASTEROIDS: USE OF OSSICLE MORPHOLOGY IN THE STUDY OF FOSSIL ASTEROIDS

The astcroids are a notoriously homoplastic group, in which strikingly similar body plans have evolved repeatedly in separate lineages, creating problems for both higher taxonomy and the reconstruction of phylogeny (Blake 1989). Examples of detailed morphological similarities between taxa seem so convincing that experienced workers have been reluctant to even admit the possibility that they are actually convergent. For example, Kesling and Strimple (1966) compared the remarkably similar abactinal disc ossicle morphology and arrangement in the Ordovician *Protopalaeaster narrawayi* Hudson, 1912 with that of the Carboniferous *Calliasterella* (see below; classified in separate orders by Spencer and Wright 1966) and concluded that convergence was most improbable. Downey (1970) was so impressed with the detailed similarities of ossicle arrangement of *Calliasterella mira* and the extant zoroasterid *Doraster constellatus* Downey, 1970 that she placed *Calliasterella mira* close to the family Zoroasteridae. This viewpoint was the basis of the polarity determination in Blake's 1987 cladogram, so such conclusions can have a long-lasting impact upon phylogenetic reconstructions.

An effective means of testing hypotheses of asteroid relationships was developed by Blake (1973), who pioneered the use of detailed morphological study of ambulacral and adambulacral ossicles in taxonomy. He used this approach to demonstrate that the extant *Luidia* (*Platasterias*) was not in fact a living somasteroid, but simply a species of the widespread and common family Luidiidae (Blake 1972, 1982; see also Madsen 1966). The ambulacral groove and mouth frame are complex structures with numerous characters that are considered to have evolved relatively slowly and therefore to provide important data on the deeper relationships of asteroids. Description and illustration of the ambulacral groove ossicles of Palaeozoic taxa (Gale 1987; Blake and Guensburg 1988; Shackleton 2005) revealed fundamental differences in construction compared with post-Palaeozoic forms. This led to the recognition of the neoasteroids as a monophyletic group (Gale 1987) and effectively falsified the hypotheses of relationship suggested by Fell (1963), Kesling and Strimple (1966) and Downey (1970).

In the present work, the process of testing hypotheses of classification and phylogenetic relationships using ossicle morphology and stereom construction is extended to the extant neoasteroids and the cladogram (Text-figs 20–21, 27) can form the basis for an assessment of homoplasy in the group. One character, which has long been used in higher classification of asteroids, is the relative size and shape of the marginal ossicles (Sladen 1889), which is still widely used as the basis for taxonomic assignment. The use of this character can be illustrated with reference to odontasterids. The description of the Odontasteridae by Clark and Downey (1992) makes a good case for monophyly of the family, because of the presence of unique synapomorphies such as unpaired, distally directed hyaline oral spines. The family contains taxa with a considerable diversity of constructional types, from forms with large blocky marginal ossicles forming a conspicuous border to the ambitus (e.g. *Odontaster mediterraneus* (Marenzeller, 1893) to those with tiny paxilliform marginals which are barely larger than the abactinals (*Odontaster validus*)). However, the taxonomic position of the family has always been informed by those taxa with large marginals, which resemble those of goniasterids and the odontasterids have therefore traditionally been placed in the Valvatida (Spencer and Wright 1966; Blake 1987). The discovery in the present study that *Odontaster validus* shares numerous synapomorphies of the adambulacrals, adambulacrals, mouth frame and stereom construction with spinulosid taxa provides another example of the extensive homoplasy of marginal morphology. The advent of molecular phylogeny is likely to throw up numerous further examples of homoplasy in overall skeletal construction, as shown by O'Loughlin and Waters (2004), who transferred the genus *Dermasterias* Perrier, 1875 from the Asteropseidae to the Asterinidae.

A further consequence of cladistic analysis of the neoasteroids is the discovery of extensive paraphyly at the level of families and higher taxa. Indeed, the difficulties encountered in adequately diagnosing extant asteroid families, and the frequent transfer of genera between these families (e.g. Clark and Downey 1992) is symptomatic of the problem. It is concluded that monophyly of asteroid families cannot be taken for granted and that investigation into ossicle morphology and molecular phylogeny is required to test relationships.

Ossicle morphology provides a valuable tool in the reconstruction of phylogenies and a reliable means of characterizing monophyletic groups and thus the identification of extant taxa. For example, the distinctive construction of the ambulacral–adambulacral contact found in all examples of the Cribellina examined by the author (a total of seven genera of Goniopectinidae and Porcellanasteridae; some illustrated, see Pl. 6, figs 4–6; Text-figs 4, 10) and similarly in all poraniids (Text-figs 8E, 11I) provides numerous additional characters for identifying respective groups. Ossicle morphology is also useful for recognizing genera and species (Blake 1973).

The value of comparative morphology of asteroid ossicles is further illustrated in the section 'Systematic Palaeontology' below, in which complete ossicle series are

described for three asteroid species from the Upper Jurassic (Oxfordian) of the French Jura. Although individuals
with ossicles in original association do occur here, much
of the material was washed from clays as isolated ossicles
and can be identified to family level by comparison with
extant taxa. The species are identified as a benthopectinid,
a pterasterid and a close relative of the zoroasterids. The
material is so well preserved that it is possible not only to
identify key synapomorphies of families (e.g. the distinctive ambulacral–adambulacral articulation of benthopectinids), but also precise relationships with other genera and
species of the family.

STRUCTURAL EVOLUTION AND
ADAPTIVE MORPHOLOGY OF THE
NEOASTEROID SKELETON

The origination of the neoasteroid clade involved important changes in the orientation, articulation and muscularization of the ambulacral groove ossicles. These
changes provided greater arm flexibility and more precise
muscular control of movements (Heddle 1967; Gale
1987). At approximately the same time, evolution took
place in the construction of the mouth frame, including
size decrease and increased muscularization of the odontophore (Gale 1987). These modifications provided a
more distensible, better-muscularized mouth structure,
enabling the ingestion of large prey (Blake 1988a, b). Dissociation of the marginals and odontophore allowed the
development of broad interradial arcs occupied by actinal
ossicles. These allowed the disc to enlarge to accommodate digestive and reproductive structures.

Once the ambulacral groove and mouth frame structures had evolved, they changed relatively little during
much of the subsequent evolutionary history of the neoasteroids. Within this basic constructional 'bauplan', most
of the evolution involved parameters of relative arm and
disc size and shape, proportionate size, number and morphology of abactinal and marginal ossicles and spine
development. The range of these variations was almost
exactly the same as found in the Palaeozoic stem group
asteroids, and homoplasy between the groups is both
common and misleading.

The data presented here support the conclusion of Gale
(1987) that the absence of flat-tipped or suckered tube
feet, the inability to feed by stomach eversion and the
absence of a brachiolarian larva are primitive features of
the Paxillosida. The presence of pointed tube feet and a
simple, presumably noneversible, stomach in exceptionally
well-preserved material of the Silurian genus *Bdellocoma*
(Sutton *et al.* 2005) supports the argument that Palaeozoic asteroids lacked these characters. The acquisition of
flat-tipped tube feet, the ability to evert the stomach and

development of brachiolaria larvae must have taken place
very early in the history of the neoasteroids, coincident
with the first occurrence of the Surculifera in the Triassic
(Text-fig. 27).

The basal neoasteroid group, the Paxillosida, comprises
specialized burrowers, and basal members of the Tripedicellaria (Goniasteridae) and Spinulosida (Odontasteridae)
include infaunal taxa. It is therefore probable that the
common ancestor of the neoasteroids was adapted to an
infaunal mode of life. Adaptations to burrowing in neoasteroids include:

- paxilliform abactinal ossicles, with short spines of
  even length providing an umbrella like cover to protect papulae and prevent sediment clogging the underlying papulae (Nichols 1969, fig. 21d);
- a flexible abactinal membrane extending to the arm
  tips to allow flexure of the ambulacral groove in the
  burrowing process (Heddle 1967);
- large marginal ossicles with intermarginal fascioles to
  provide abactinal–actinal respiratory pumps (Shick
  et al. 1981). The fascioles may extend across the actinal interareas;
- pedicellariae restricted to the ambulacral groove, usually attached to adambulacral ossicles;
- presence of additional muscles running between marginals, adambulacrals and supramambulacral ossicles
  (Heddle 1967).

The evolution of flat and suckered tube feet enabled the
Surculifera to adapt to an epifaunal existence on rocky
substrata and probably coincided with the development
of extraoral feeding (Gale 1987). From the cladogram
(Text-figs 20, 27), it appears that the neoasteroids developed an epifaunal mode of life at least three times. Benthopectinids, such as *Benthopecten simplex*, which possess
long abactinal and marginal spines are presumably secondarily epifaunal. The goniasterids are basal Surculifera
and include forms adapted to burrowing and others to an
epifaunal existence; likewise, the basal spinulosids, the
Odontasteridae, include taxa that are burrowers and surface dwellers. Neoasteroid evolution therefore contrasts
strikingly with that of post-Palaeozoic echinoids, which
were plesiomorphically epifaunal and only subsequently
developed infaunal adaptations (Smith 1990).

Adaptations to epifaunal life habits in the neoasteroids
include:

- domed abactinal surfaces with enlarged primary and
  radial ossicles, for increased protection;
- more effective muscularized alveolar and pedunculate
  pedicellariae for protection from larval settling and
  parasites;
- diverse forms of abactinal and marginal spines,
  including a dense cover of flat-granular and short-

blunt spines over entire body surface or sharp thorny spines, either alone or in brush-like clusters;
– areas of smooth dense stereom on external ossicles.

Carnivorous specialists (e.g. Asteriidae, Solasteridae) developed lighter, more flexible reticulate skeletons, imbricate marginal and abactinal ossicles to allow flexure over prey species (Eylers 1976), and some asteriids used wreaths of abactinal crossed pedicellariae to catch prey (Chia and Amerongen 1975).

Only four neoasteroid lineages have evolved entirely novel morphological innovations, unknown elsewhere in the previous history of the Asteroidea; the Cribellina (Ctenodiscidae, Porcellanasteridae), the Brisingida (Freyellidae Downey, 1986; Brisingidae G. O. Sars, 1875), the Velatida (Pterasteridae and, to a lesser extent, Korethrasteridae) and the Xyloplacidae Baker, Rowe and Clark, 1986. The porcellanasterids (Madsen 1961) are paxillosids in which the intermarginal fascioles have evolved into highly efficient lamellar respiratory pumps called cribriform organs (Schick *et al.* 1981). The mouth frame has developed elongated circumorals that imbricate distally, probably increasing the distensibility of the structure. Segmental organs, with a flap or valve formed from modified adambulacral spines, are also present. Their function is unknown (Madsen 1961; Clark and Downey 1992, fig. 3l).

The Brisingida comprise multiarmed forms with a small, rounded disc, well demarcated from the long narrow arms, superficially closely resembling ophiuroids. The mouth frame ossicles form a perfectly circular, narrow ring and the ossicles are immobile, with the odontophore fused to the orals in *Brisinga costata*. The ossicles of the ambulacral groove are also modified, such that the groove cannot be opened or closed and the internally positioned abactinal transverse ambulacral muscles and interdigitating dentition effectively lock the cross-groove articulation in place.

The pterasterids are a highly derived group of neoasteroids, and one species of the family (*Pteraster tesselatus* Ives, 1888) has a pattern of coelomogenesis unique amongst asteroids (Janies and McEdward 1992). The novel structures found in the family include:

– specialized adambulacrals with a lateral extension, with an additional muscle and articulation surface and a large specialized abactinolateral spine;

– interradial grooves floored with imbricating chevron plates;
– an abactinal membraneous cavity, opening to the exterior via a central abactinal osculum and numerous smaller spiraculae and supported by modified abactinal ossicles;
– an abactinolateral membrane, running between the abactinolateral spines and forming an actinal extension to the abactinal cavity;
– specialized operculae between the bases of the actinolateral spines, each with a specialized guard ossicle, the apertural spines.

Although originally placed in a new class, the Concentricocycloidea (Baker *et al.* 1986), *Xyloplax* is now widely recognized as an asteroid, albeit a highly derived one (e.g. Smith 1988; Janies and Mooi 1998; Janies 2001). Most authors have placed the taxon in the neoasteroid clade (Smith 1988; Janies and Mooi 1998), but Mah (2006) proposed that the Concentricocycloidea were an infraclass of the Asteroidea, of equal status with the Neoasteroidea. These tiny, flattened, circular animals live on wood in the deep sea and have a very wide peristome, around the perimeter of which the modified ambulacrals and adambulacrals are arranged in a single column. The circumoral bars of the orals and circumorals are splayed out and the ambulacral heads abut these elongated ossicles – the midradial contact between ambulacral heads of opposing sides of the column has been lost. However, the skeletal construction of *Xyloplax* has never been described in detail; the only description illustrated with SEM photographs of ossicles (Rowe *et al.* 1988) contains errors of interpretation of ossicle identity and homology. Thus, their 'adambulacrals' (Table 7; pl. 4, figs 56–57) are clearly ambulacrals, and the 'marginals' (pl. 4, figs 58–61) are adambulacrals. They figured, as mouth frame ossicles, the oral ossicle (pl. 4, fig. 52) and circumorals (pl. 4, figs 53–54) and the odontophore. Some of these mistakes were corrected by Janies and Mooi (1998), but both these authors and Mah (2006) incorrectly identified the circumoral as an oral ossicle. Although only limited material of *Xyloplax* was available during the present study, the reinterpretation of the neoasteroid skeleton made in this paper provides a template for the assessment of homologies and relationships of this highly derived asteroid.

TABLE 7. Homologies of skeletal elements in *Xyloplax*.

Rowe *et al.* (1988)	Smith (1988)	Janies and Mooi (1998)	Mah (2007)	Gale, this paper
Superomarginal (Pl. 4 fig 43)	Abactinals	Abactinals	Abactinals	Abactinals
Mouth frame ossicle (Pl. 4, fig. 52)	Oral	Odontophore	–	Oral
Mouth frame ossicle (Pl. 4, fig. 53,4)	–	Oral	Oral	Circumoral
Inferomarginal (Pl. 4, figs. 58–62)	Adambulacral	Adambulacral	Adambulacral	Adambulacral
Adambulacral (Pl. 4 figs 56,7)	Ambulacral	Ambulacral	Ambulacral	Ambulacral
Odontophore	–	–	–	Odontophore

Primarily, *Xyloplax* is clearly a neoasteroid because it has an internal odontophore that articulates with the interior of the distal oral ossicle (character 97, 0–1; see Rowe *et al.* 1988, pl. 4, fig. 55). There have been several suggestions of the relationships of *Xyloplax* in other neoasteroids. First, Smith (1988) suggested that *Xyloplax* was sister taxon to the deep-water genus *Caymanostella* Belyaev, 1974, based on the comparable overall shape, wide peristome, flattened, tesselate abactinals and the madreporite that forms a simple pore. As a consequence, Belyaev (1990) proposed the group Peripodoidea to include both *Caymanostella* and *Xyloplax*. Janies and Mooi (1998) found *Xyloplax* to be most closely related to the forciculatid *Rathbunaster* Fisher, 1906 from molecular evidence (18S and 28S rDNA), and Janies (2001) argued, on the basis of molecular data, that *Xyloplax* was most closely related to *Pteraster*. It is now possible to test these hypotheses using detailed morphological analysis of the mouth frame and other ossicles.

The oral ossicles of *Xyloplax turnerae* Rowe, Baker and Clark, 1986 (Text-fig. 33D, F) are similar to those of other neoasteroids (compare Text-figs 16, 17) and possess the usual structures for articulation with the odontophore and a notch for the ring nerve (*rng*). However, the apophyse is very elongated, and the ring vessel groove is partly internal. The circumorals (Text-fig. 33D, E) are elongated and display two articulation structures: one elongated and running along the length of the ossicle, the other short and positioned at the interradial end of the ossicle. The elongated articulation is interpreted as homologous with the distal circumoral articulation (*dcoa*; Text-fig. 19), the short one with the proximal circumoral articulation (*pcoa*). The circumoral heads meet across the radius (Text-fig. 33B). The odontophore (Text-fig. 33H) is short and broad, with clear articulation surfaces for the interradial face of the oral. The first adambulacral articulates with the radial surface of the oral ossicle (Text-fig. 33B).

TEXT-FIG. 33. Skeletal construction and ossicle homologies in Xyloplacidae. A, *Ankyloplax janetae*, based on topotypical material, Gorda Ridge, north-east Pacific, see Mah (2006) for details. B, D–F, H, *Xyloplax turnerae*, western Atlantic, Bahamas, redrawn from published illustrations. B, construction of an interradial-radial segment to illustrate interpretation favoured in the present paper. D, actinal and E, abactinal aspects of circumoral ossicle, after Rowe *et al.* (1988, pl. 4, figs 53–54). F, oral ossicle in abactinal (interradial) view, after Rowe *et al.* (1988, pl. 4, fig. 52). H, odontophore, abactinal view, after Rowe *et al.* (1988, pl. 4, fig. 55). C, G, I, *Caymanostella spinimarginata*, isolated ossicles (Table 1). C, circumoral ossicle, in abactinal aspect. G, oral ossicle in abactinal (interradial) aspect. I, odontophore, in abactinal aspect. A, B, ×45; C, G, I, ×40; D–F, H, ×70.

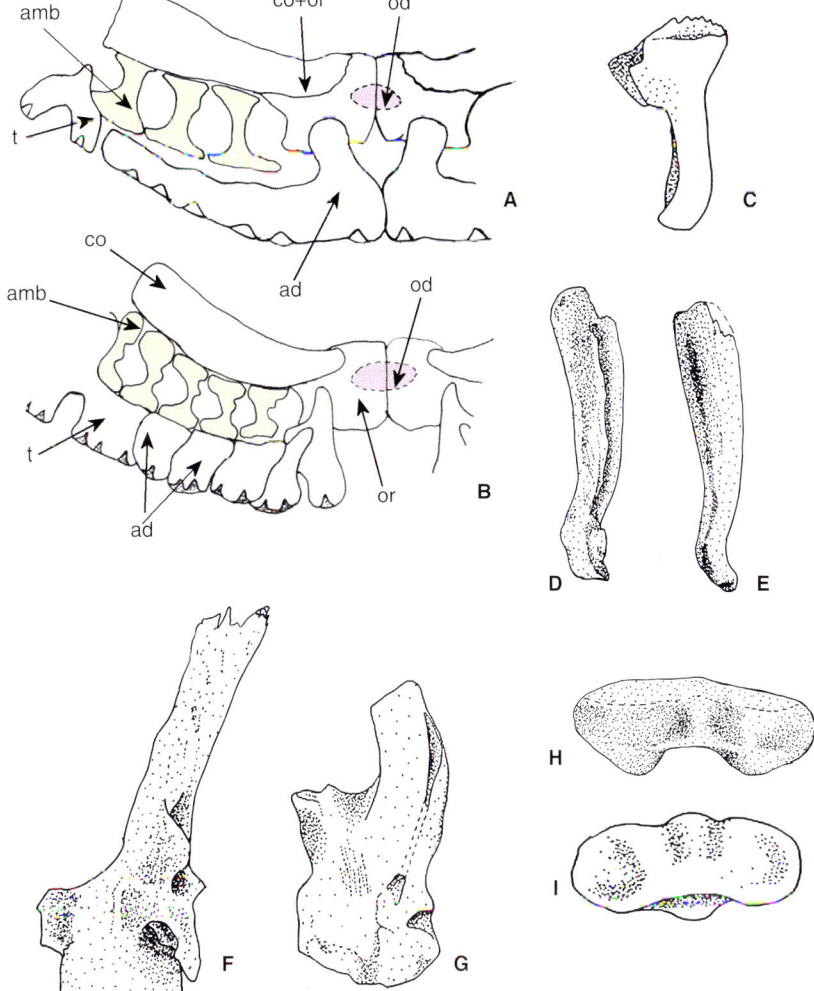

The basic construction and homologies of the mouth frame and its constituent ossicles are therefore similar to those of every other neoasteroid. The most important morphological differences between *Xyloplax* and other neoasteroids lie in: (1) the complete loss of muscularization of the groove and mouth frame ossicles; (2) the extreme elongation of the oral apophyse and circumoral ossicles and their rotation to a margin-parallel position; and (3) separation of the ambulacral rows of a single groove by loss of the adradial contact, such that the ambulacrals lie parallel to the margin of the asteroid and are set between the adambulacrals and the circumorals.

The ossicles of the mouth frame of *Xyloplax turnerae* show some resemblances to those of *Caymanostella spinimarginata* Belyaev, 1974. Although the oral ossicles differ on shape, the apophyse of the oral is elongated in this species as well, and the ring vessel groove is also partly internal as in *Xyloplax turnerae* (Text-fig. 33G). The odontophore is closely similar in morphology in the two taxa, with a short, broad oval form (Text-fig. 33H, I). The distal bar of the circumoral is very elongated in *Caymanostella spinimarginata*, but the articulation surfaces for the oral (*dcoa*, *pcoa*) are small as in all other neoasteroids. On the strength of this evidence, in addition to the characters listed by Smith (1988), *Xyloplax* would appear to be most closely related to *Caymanostella*. However, this requires further investigation.

The above description and comparison apply to *Xyloplax medusiformis* Baker, Rowe and Clark, 1986 and *Xyloplax turnerae*. However, the species described as *Xyloplax janetae* Mah, 2006, has a highly modified construction (Text-fig. 33A) and is here placed in a new genus, *Ankyloplax* (see section 'Systematic Palaeontology'). The oral and circumoral are fused to form a single, bar-shaped ossicle, and the single adambulacral ossicle is greatly elongated to form a curved margin to the disc that carries six to eight spines along its lateral margin. The ossicle resembles the blade of a scythe and a process articulates interradially with the basal part (radial surface) of the oral ossicle. It is not clear whether this plate arises from fusion of a number of adambulacrals, or is formed by extended growth of a single ossicle. Three ambulacrals are present in each half radius, and these do not articulate with the adambulacral, but are separated by a narrow zone floored by the interior of the abactinal ossicles. Thus, in *Ankyloplax janetae*, each half radius is composed of three ambulacrals, one adambulacral and a single fused oral-circumoral and shares a terminal and an odontophore with adjacent half radii. This is without doubt the most extreme modification of the neoasteroid skeleton hitherto described.

It is noteworthy that these four families are either exclusively or dominantly deep-sea groups and that many genera or even species of the families are of cosmopolitan distribution. They appear to have adapted to different deep-sea resources, as the brisingids are suspension feeders, the porcellanasterids are deposit feeders, the pterasterids predators and scavengers and *Xyloplax* feeds on decaying wood. However, there is no evidence that families such as the zoroasterids, benthopectinids and pterasterids occupied the deep sea prior to the Oligocene, and these families (or their close relatives) occur commonly in shallow-water settings in the Jurassic and Cretaceous (see below section 'Systematic Palaeontology'). Zoroasterids occur in shallow-marine sediments in the Eocene of Antarctica (Blake and Zinsmeister 1979). It therefore appears that the occupation of the deep sea by specialized asteroid families did not commence until after the development of cold ocean bottom waters in the late Paleogene.

SYSTEMATIC PALAEONTOLOGY

Genus CALLIASTERELLA Schuchert, 1914

Calliasterella mira (Trautschold, 1879)
Plate 17; Text-figures 5A, 6, 7B–D, 14, 15B, 34

1879 *Calliaster mirus* Trautschold, p. 10, pl. 2, figs 3–4.
v. 1909 *Calliaster mirus*; Schöndorf, p. 327, pl. 23, figs 2–5; pl. 24, figs 1–18.
1914 *Calliasterella mira*; Schuchert, pp. 190–191, fig. 11.
1970 *Calliasterella mira*; Downey, pp. 2, 4, figs 1, 2.

EXPLANATION OF PLATE 17

Photographs of the specimen of *Calliasterella mira* described by Schöndorf (1909) from the collections of PIN, Moscow (unregistered), Mosquensis Horizon, Upper Carboniferous (Moscovian), Moscow Basin.

Figs 1–2. Entire specimen, showing both sides. The side with dissociated ossicles (1) was probably uppermost in the sediment.

Figs 3–6. Cleaned ossicles of one interradius of the mouth frame, comprising the axillary/odontophore and pair of oral ossicles, in actinal (3), internal (4), distal (5) and external (6) aspect.

Fig. 7. Enlarged adambulacral ossicles to show details of articular surfaces and spine attachments.

Fig. 8. Details of adambulacral and marginal rows to show nature of adambulacral imbrication. See also Text-figures 6, 7 and 14 for comparison.

Scale bars for Figs 1–6 represent 10 mm and 5 mm for Figs 7, 8.

PLATE 17

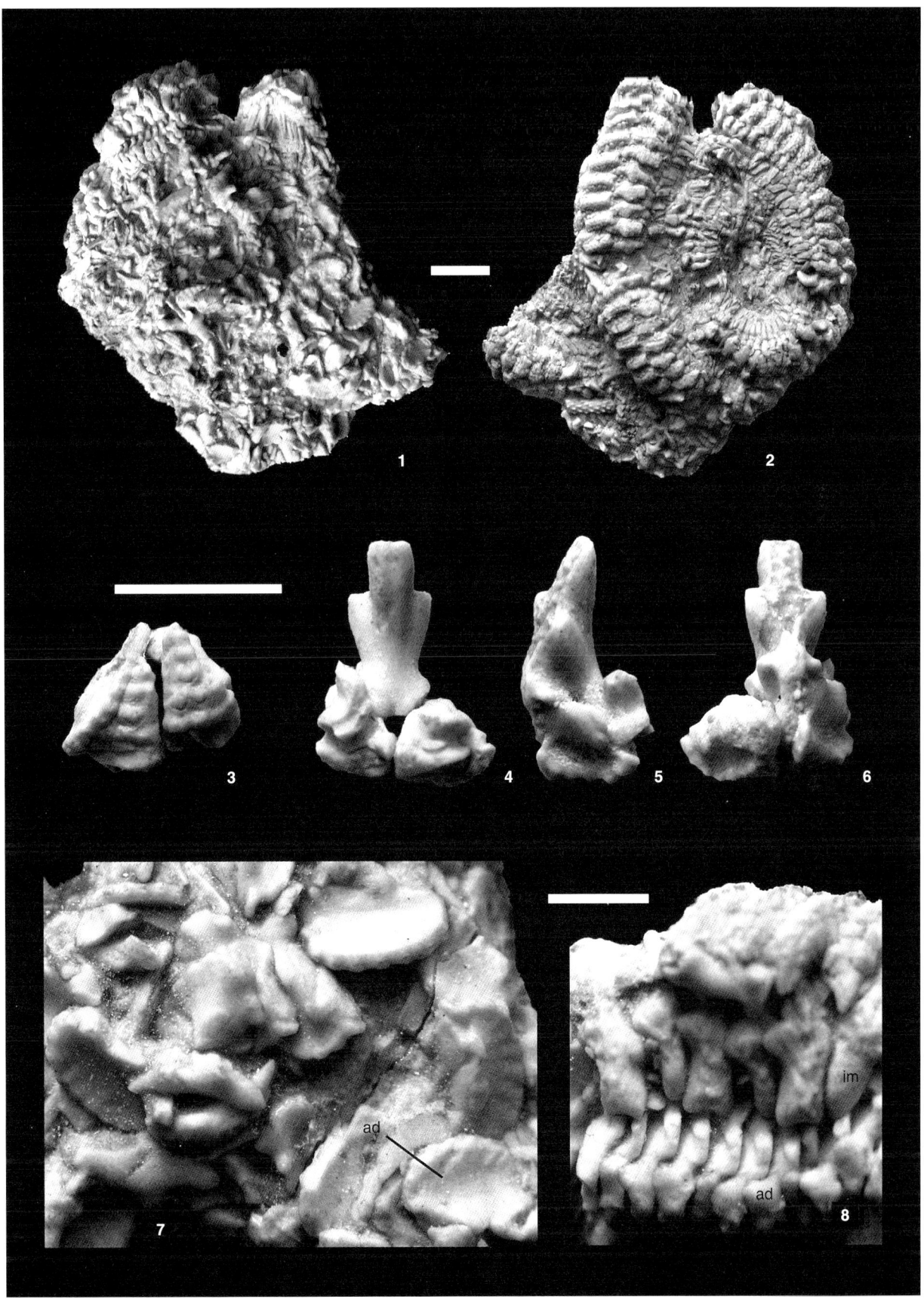

GALE, *Calliasterella mira*, Moscovian of Russia

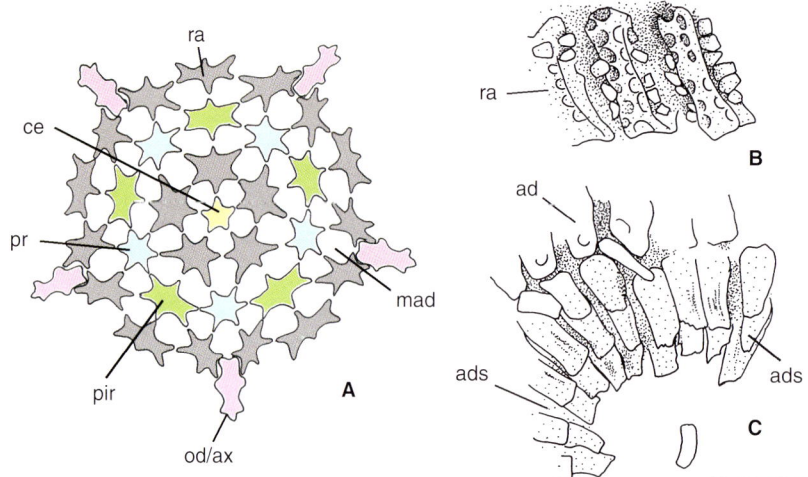

TEXT-FIG. 34. *Calliasterella mira.* A, reconstruction of abactinal plating of disc, after Schöndorf (1909). B, radial ossicles, to illustrate the symmetrically arranged spine pits and short, flattened spines along proximal and distal margins. C, adambulacrals and adambulacral spines in lateral view, of flattened, imbricating form. Scale bar represents 2 mm.

Holotype. The whereabouts of the fragmentary holotype specimen figured by Trautschold are unknown. The more complete individual illustrated by Schöndorf (1909) is now (unregistered) in the collections of PIN, Moscow. Both are from the Mosquensis Bed at Miatschkowa near Moscow (Upper Carboniferous, Moscovian).

Description. The lower surface of the type specimen (Pl. 17, fig. 2) comprises parts of the disc and two arms, essentially articulated, and the flat upper surface (Pl. 17, fig. 1) includes an arm fragment and numerous well-preserved dissociated ossicles, most conspicuously adambulacrals. This type of preservation has been described in Jurassic isocrinid crinoids and attributed to disturbance of the upper surface on the sea floor by current action. The abactinal ossicles of the disc illustrated in Schöndorf's photograph (his pl. 23, figs 4–5) and accompanying drawings have been removed from the specimen, possibly by Schöndorf himself, and are presumably now lost. The axillary ossicle visible in Schöndorf's photograph and the associated pair of orals have been prepared out from the specimen. Re-examination of the specimen confirms the accuracy of Schöndorf's observations, the precision of his illustrations and the likelihood that his reconstructions are faithful and serves to dispel the doubts cast on the work by Downey (1970).

The disc was of moderate size and gently domed (Schöndorf 1909, pl. 23, figs 4–5), the interbrachial arcs rounded. The abactinal ossicles of the disc are rather few and large, with a stellate outline, and are thin in cross-section (Schöndorf 1909, pl. 24, figs 1–9). Following Schöndorf's reconstruction (Text-fig. 34A), a pentagonal centrale is surrounded by five large primary radial ossicles, which are in turn surrounded by a circlet of five smaller primary interradials. Paired interradial ossicles are present distal to the primary interradials. Radial plates 2 and 3 are enlarged in comparison with more distal radials of the arm. The axillary (Pl. 17, figs 4–6; Text-fig. 14) is tall and short and articulates with the first two inferomarginals, the paired interradials and the oral ossicles below.

Although incomplete, the arms are elongated and subcylindrical in cross-section (Text-fig. 1A), with R:r probably in excess of 10:1. They are made up of a single broad row of rectangular radial ossicles, two rows of inferomarginals and two rows of broad, short adambulacrals (Pl. 17, fig. 2). The ambulacrals are proportionately narrow and rather small and inconspicuous. The external margins of abactinals, marginals and the axillary ossicles carry a single row of short, broad spines that were attached to notched articulation surfaces (Text-fig. 34B). The radials are rectangular, 3–4 times broader than long and alternate with the inferomarginals. They carry a row of 5–6 short, flattened spines along both the proximal and distal margins (Text-fig. 34B). The spines attach to notches in the surface of the radial ossicle. The inferomarginals have a bifid abactinal process which articulates with two radial ossicles and short proximal and distal articulation processes articulate with adjacent inferomarginals of the same row. One to three adambulacrals correspond with a single inferomarginal.

The ambulacrals are robust and equilaterally triangular in proximal and distal aspect with a concave abactinal margin (Text-fig. 6D–G). The ossicles are differentiated into discrete head, waist and basal regions and are notched for passage of an internal ampulla. The heads of opposing ambulacrals articulated across the groove by means of an interdigitating dentition, made up of short ridges and grooves parallel with the length of the arm. Proximal and distal attachment sites for the longitudinal interambulacral muscles and articulation surfaces are present on the ambulacral head (Text-fig. 6D–G). There are no surfaces for insertion of the transverse ambulacral muscles which were completely absent. Ambulacrals and adambulacrals alternate. The ambulacral base carries a central transverse ridge for articulation with the distal part of the proximal adambulacral (*ada1*). This terminates adradially in an enlarged boss for articulation with the proximal adambulacral (*ada2*) and wing-like proximal and distal extensions for insertion of the proximal and distal ambulacral–adambulacral muscles. The wings are asymmetrical, with an enlarged distal (abradial) extension.

The adambulacrals are short, broad and oval in proximal/distal aspect and imbricate slightly distally with adjacent ossicles (Pl. 17, figs 7–8; Text-fig. 6A–C). The flattened abactinal surface carries structures that articulate with the ambulacrals and inferomarginals, and two asymmetrical facets for attachment of proximal and distal adambulacral–ambulacral muscles are visible. The larger proximal muscle attachment shows several concentric growth surfaces. Two prominent interadambulacral articulation structures are present. The proximal surfaces are gently convex, the distal pair flat or slightly concave and are constructed of smooth imperforate stereom. The abradial articulation is transversely elongated, the adradial one round. The attachment site for the adambulacral–adambulacral muscle is large and distinct. The actinal margin of each adambulacral carries eight spines, set in a single transverse row and borne on large attachment sites. The adambulacral spines are short and transversely flattened, with a concavo-convex cross-section (Text-fig. 34C).

The mouth frame and its constituent ossicles are very well preserved and developed free of matrix (Pl. 17, figs 4–6; Text-figs 14, 15B). The oral ossicles are flattened actinally and carry four rows of spine attachments. The first adambulacral is very broad and short and occupies the entire width of the radial/distal surface of the oral ossicle. Prominent paired oral-adambulacral articulation structures are present, and the insertion site of the oral-adambulacral muscle is well defined. The apophyse of the oral was an attachment for radial and interradial interoral muscles and articulation with the proximal circumoral bar. The ring nerve and ring vessel grooves are clearly visible in proximal view of the orals.

The axillary/odontophore is situated between the oral ossicles and has a tall, narrow external surface, from either side of which short symmetrical processes extend to contact the proximal surfaces of the first inferomarginals and the interradial faces of the oral ossicles. The external face of the axillary (Text-fig. 14A) carried short spines attached to shallow notches. The nature of the oral-axillary contact is well exposed, and homologous structures to the proximal and distal jaws of the neoasteroid odontophore can be identified, as can the keel (Text-fig. 15). The keel meets an articulation ridge on the inner face of the oral, and a depression actinal to this ridge was a likely attachment site of the homologue of the oral-odontophore muscle of neoasteroids.

Although now lost, the circumoral ossicle was figured by Schöndorf (1909, pl. 24, figs 15–16). This is low, with an elongated head and the short proximal bar that contacted an elongated proximal circumoral bar on the apophyse of the oral. There were no intercircumoral muscles.

Discussion. Kesling and Strimple (1966) described *Calliasterella americana* from the Pennsylvanian of Illinois, a small asteroid in which the arrangement of the abactinal plating and development of the marginals is comparable with *Calliasterella mira,* although these ossicles are much more robust in the former. However, the construction of the adambulacrals is completely different in the two species. In *Calliasterella mira,* these are short and broad and carry a transverse row of up to eight spines, whereas

in *Calliasterella americana* these are narrow, curved and imbricate; the adambulacral spines are concentrated on the furrow, and each plate bears a cluster of spines of varying sizes (Kesling and Strimple 1966, pl. 153, figs 3–4). These are important differences, and the American species requires a new genus.

The suggestion by Downey (1970) that *Calliasterella mira* falls close to the extant Zoroasteridae can now be refuted from a detailed description of the ossicle morphology of the ambulacral column and mouth frame. Although the ambulacral and adambulacral ossicles of *Calliasterella mira* show various characters found in the crown group neoasteroids (e.g. alternate ambulacrals–adambulacrals, presence of ambulacral–adambulacral articulation surfaces and muscles), the mouth frame is primitive, with a large axillary that articulates with the inferomarginals. The striking similarity between the zoroasterid *Doraster constellatus* and *Calliasterella mira* is therefore convergent.

Subclass NEOASTEROIDEA Gale, 1987
Clade PAXILLOSIDA Perrier, 1884
Clade CRIBELLINA Fisher, 1912
Family GONIOPECTINIDAE Verrill, 1899

Genus CHRISPAULIA Gale, 2005

Type species. Nymphaster radiatus Spencer, 1905, by original designation.

Diagnosis. Goniopectinids that possess blocky margins with a sculpture of small bead-like bosses on the central areas.

Chrispaulia jurassica sp. nov.
Plate 18, figures 1–11

Derivation of name. With reference to the Jurassic age of the species.

Holotype. NHM EE13586 (Pl. 18, figs 2–3); superomarginal, from the upper Oxfordian marls (*bifurcatus* Zone, *stenocycloides* Subzone) at Savigna, near Orgelet, Jura, France (Hess 1971).

Paratypes. The illustrated ossicles (Pl. 18, figs 1, 3–4, 6–9) (NHM EE 13585, 13587–13591), from the upper Oxfordian marls (*bifurcatus* Zone, *stenocycloides* Subzone) at Savigna, near Orgelet, Jura, France (Hess 1971).

Material. Fifty-four marginal ossicles from Savigna; two marginals and one adambulacral ossicle from Terres Rouges, Swiss Jura (Hess 1971).

Diagnosis. Chrispaulia in which the central areas on the marginals are poorly defined; the inferomarginals possess a central cluster of larger spine pits.

Description. The marginals are nearly cuboidal in form, with well-demarcated lateral and actinal/abactinal surfaces set at right angles to each other (Pl. 18, figs 1–9). A central area made up of rugosities is present on the inferomarginals, and at the distal end of this a group of two or three large, bifid spine bases occur (Pl. 18, figs 4–5, 7–8). Distal marginals are proportionately more elongated. The superomarginals carry a sculpture of rugosities which coalesce to form irregular transverse ridges on the abactinal–lateral region of the ossicle (Pl. 18, figs 1–3, 6, 9). The margin of the outer faces is made up of fine, even rugosities that bore spines of the cribriform organs in life. The boundaries between the central regions and margins of the ossicles are poorly demarcated. A single adambulacral (Pl. 18, figs 10–11) from Terres Rouges in the Swiss Jura (see Hess 1971) can be assigned to this species with reference to the more extensive material of *Chrispaulia radiatus* (see Gale 2005; Pl. 18, figs 12–13 herein). This is elongated and carries a low adradial prominence. Several subadambulacral spine bases are present, and a row of small furrow spines bases occurs on the adradial surface of the outer face. The inner (abactinal) face of the ossicle provides evidence of the goniopectinid affinity of this specimen, including the small, single, rugose *ada1*, the confluent *adad/ada2*, and the elongated *ada3* (compare Text-fig. 10E,F; Gale 2005, fig. 9A–C).

Remarks. This species differs from *Chrispaulia radiatus* in the more irregular, poorly demarcated outline of the central area on the ossicles and the apparent absence of attachment surfaces for fasciolar cover spines.

Family BENTHOPECTINIDAE Verrill, 1899

Genus JURAPECTEN gen. nov.

Derivation of name. With reference to the type locality in the French and Swiss Jura.

Type species. Jurapecten hessi sp. nov.

Diagnosis. A benthopectinid that lacks transverse abactinal ridges on the ambulacral ossicles and possesses elongated, bar-like ambulacral heads. Parapaxillae are absent.

Jurapecten hessi sp. nov.
Plate 19, figures 112, 15–17; Plate 20, figures 5–8, 10, 12–14, 16, 19

Derivation of name. In honour of Hans Hess, who discovered the echinoderm fauna of Savigna and Terres Rouges.

Holotype. NHM EE 13594 (Pl. 19, fig. 5); a cluster of ossicles, from the upper Oxfordian marls (*bifurcatus* Zone, *stenocycloides* Subzone) at Savigna, near Orgelet, Jura, France (Hess 1971).

Paratypes. NHM EE 13592, 13593, 13595–13609 (Pl. 19, figs 1–12, 15–17; Pl. 20, figs 5–8, 10, 12–14, 16, 19), from the upper Oxfordian marls (*bifurcatus* Zone, *stenocycloides* Subzone) at Savigna, near Orgelet, Jura, France (Hess 1971).

Material. Five groups of associated ossicles and spines, including ambulacrals, adambulacrals, abactinals, terminals and spines, plus over 1000 isolated ossicles including marginals, orals, circumorals, adambulacrals, ambulacrals and terminals.

Diagnosis. As for genus.

Description. No specimens showing the form of the body are known, but from the close similarity of ossicle types with those of extant species, it probably had a small disc and long, proportionately narrow arms. All ossicles are small, and extant specimens (e.g. *Cheiraster* sp. juv., Indian Ocean; Table 1) have ossicles of comparable size at R of about 10–20 mm. However, the fine, dense, well-differentiated stereom of the Savigna material indicates that the specimens were adult. The adambulacrals (Pl. 19, figs 11–12) are rectangular in abactinal and actinal aspect, and the actinal surface carries a single rounded, bifid

---

PLATE 18

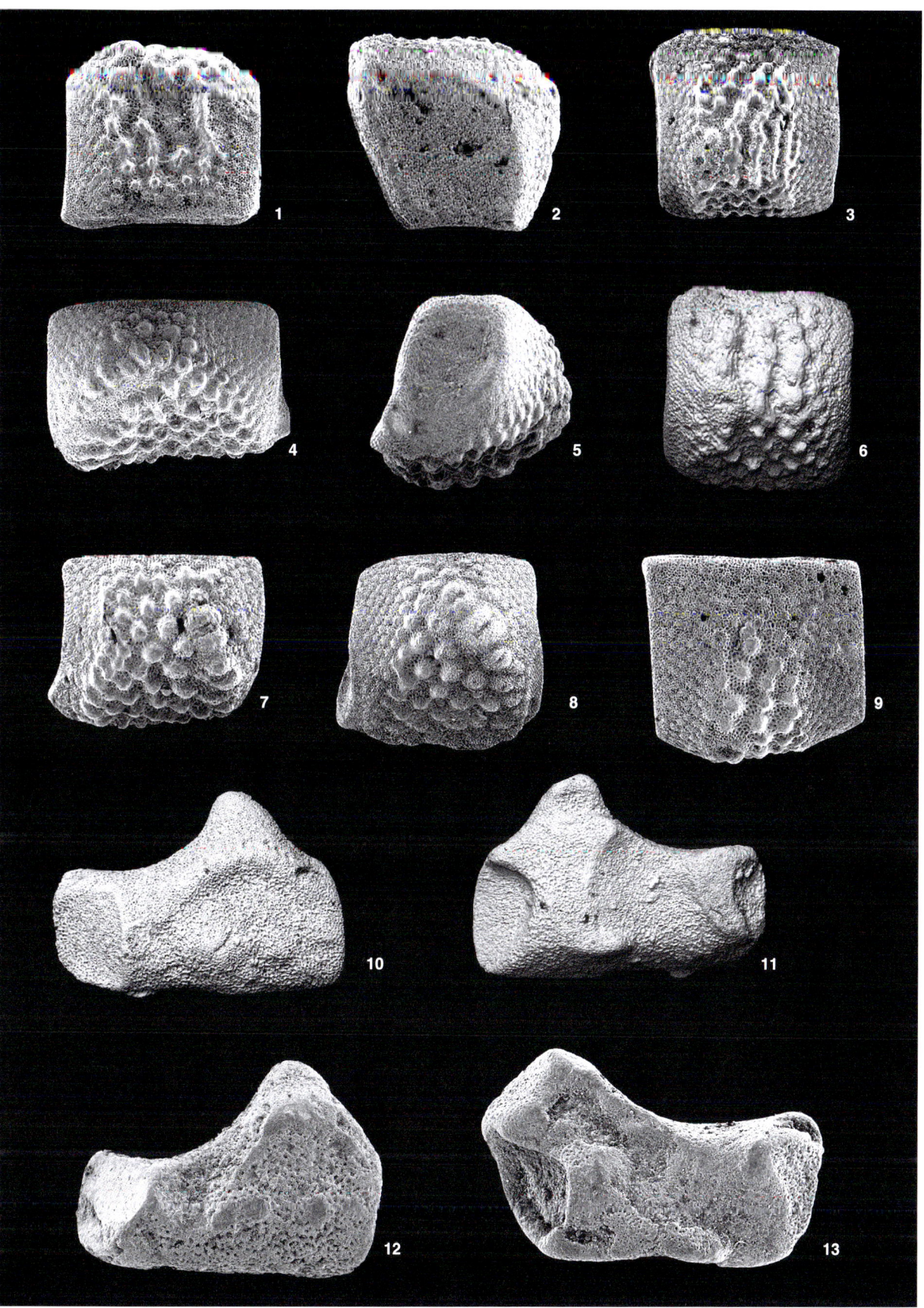

GALE, Fossil gonioipectinids

base for a subadambulacral spine set on an oblique ridge. The adradial prominence (Text-fig. 4) carries 4–6 notches for attachment of furrow spines. The abactinal (internal) face bears an adradial ridge on which *adad* and *ada2* are positioned. *Ada3* is present on the abradial distal margin of the ossicle. All features of the adambulacrals are closely comparable with those displayed by extant benthopectinids. The ambulacrals are hourglass-shaped in actinal/abactinal aspect (Pl. 19, figs 8–9), and the ambulacral head is an elongated bar forming the longest part of the ossicle. The actinal surface of the ambulacral base is rectangular and carries an oblique ridge for articulation with the adambulacral (*ada2*) and strongly asymmetrical sites for attachment of *dadam* and *padam*, similar to those in *Cheiraster gazellae* Studer, 1883 (Pl. 19, figs 13–14). In proximal/distal aspect (Pl. 19, fig. 10), the ambulacrals are triangular and lack the strongly raised transverse central ridge and articulation facet for inferomarginal contact present in all extant benthopectinids (Pl. 19, figs 14–15). Infero- and superomarginals alternate (see Pl. 19, fig. 1), such that marginals of a single row contact two ossicles of the adjacent row. Superomarginals are longer than high, narrow, with large, gently convex lateral faces (Pl. 20, figs 6–7). A single, crater-like spine pit is present on the abactinal–lateral margin of each plate, and the outer surface of the ossicles is covered by rounded rugosities. The inferomarginals are rectangular in outline, have discrete actinal and lateral surfaces set at right angles and bear a central cluster of 2–4 crater-like spine bases (Pl. 20, figs 5, 12, 16). Distal inferomarginals taper strongly towards the arm tip. The inter-inferomarginal articular surfaces bear an elongated strip of smooth, dense stereom on the actinal–lateral part (Pl. 20, figs 8–9). Abactinal ossicles are present in the associated clusters of plates; these are oval, thick and loaf-like, lacking obvious spine bases and differentiated central areas (Pl. 19, figs 3, 6). The oral ossicles (Pl. 20, figs 10, 14) are triangular in outline, with a tall stout apophyse, an elongated *dcoa* (see Text-fig. 13) and a moderately elongated, triangular proximal blade. The radial (outer) surface is gently concave and carries 2–3 crater-like bases for suboral spines, similar in shape to those on the adambulacrals for subadambulacral spines. The first adambulacral articulation surface is rhomboid. The interradial (inner) face of the oral displays a clear groove for articulation with the proximal odontophore (*poda*) and rugose stereom for interoral articulation on the actinal part of the proximal blade. Circumoral ossicles (Pl. 20, fig. 13) are typically paxillosid (compare Pl. 20, figs 11, 15; Text-fig. 19L–Q) and possess short *pcp* and elongated, distinctly angled *dcp*. Terminal ossicles (Pl. 20, fig. 19) are triangular with a deep, central proximal notch and are similar to terminals of *Cheiraster* Studer, 1883 (Pl. 20, fig. 18). Spines are associated with adambulacral and ambulacral ossicles (Pl. 19, figs 1–6). These are stout and conical and irregular, made up of distally diverging irregular ridges of stereom.

Remarks. The well-preserved benthopectinid material from Savigna allows direct comparison with ossicles of extant species. The highly specialized nature of the ambulacral–adambulacral articulation (elongated, flat articulation surface on ambulacral base; concave (*amb*)-convex (*adamb*) ridge carrying *ada2*; strongly asymmetrical *dadam*, *padam*) in *Jurapecten hessi* is essentially identical with that developed in modern benthopectinids (compare Pl. 19, figs 8–17). However, the ambulacrals are distinctly different, in the possession of elongated, bar-like ambulacral heads (Pl. 19, figs 5, 8–9) and the absence of the raised transverse ridge on the abactinal surface of the ambulacral shaft and base (Pl. 19, fig. 10). The absence of this structure means that there is no ambulacral articulation surface with the inferomarginal and that the longitudinal muscles of extant benthopectinids (which attach to the ridge) cannot have been present. Additionally, true parapaxillae (abactinal ossicles which bear a distinct, spine-bearing, raised central area) are not developed. The marginal ossicles are very similar to those of extant *Pontaster* Sladen, 1885 and *Cheiraster*, in the alternation of supero- and inferomarginals, shape, articular surfaces (specialized smooth stereom ridge on the inter-inferomar-

---

EXPLANATION OF PLATE 19

Figs 1–12. Associated groups of ossicles and dissociated ossicles of the benthopectinid *Jurapecten hessi* gen. et sp. nov., upper Oxfordian marls (*bifurcatus* Zone, *stenocycloides* Subzone) at Savigna, near Orgelet, Jura, France.

Figs 1, 3–4. Group of associated ambulacrals and spines (paratype, NHM EE13592).

Figs 2, 6. Group of associated ambulacrals and adambulacrals (paratype, NHM EE 13593).

Fig. 5. Associated ambulacrals and adambulacrals (holotype, NHM EE 13594).

Fig. 7. Superomarginal spine of *Benthopecten simplex* in lateral view (compare with spine in Fig. 4).

Figs 8–10. Ambulacral ossicles in actinal (8; paratype, NHM EE 13595), abactinal (9; paratype, NHM EE 13596) and proximal views (10; paratype, NHM EE 13597).

Figs 11–12. Adambulacral ossicles in actinal (11; paratype, NHM EE 13598) and abactinal (12; paratype, NHM EE 13599) aspects.

Figs 13–17. *Cheiraster gazellae*, Recent, Philippines (Table 1), ambulacrals in actinal (13), abactinal (14) and distal (15) views. Adambulacrals in actinal (Fig. 16) and abactinal (Fig. 17) views. Note the presence of abradial ridge (*abr*) on ambulacrals that articulates with inferomarginals and acts as an insertion site for longitudinal arm muscles in extant benthopectinids; the *abr* is absent in *J. hessi* (compare Figs 10 and 15).

Fig. 1, ×12; Figs 2, 15, ×15; Figs 3, 6, 13–14, 16–17, ×20; Fig. 5, ×25; Fig. 4, ×30; Fig. 7, ×20; Figs 8, 10, ×35; Figs 9, 11–12, ×40.

PLATE 19

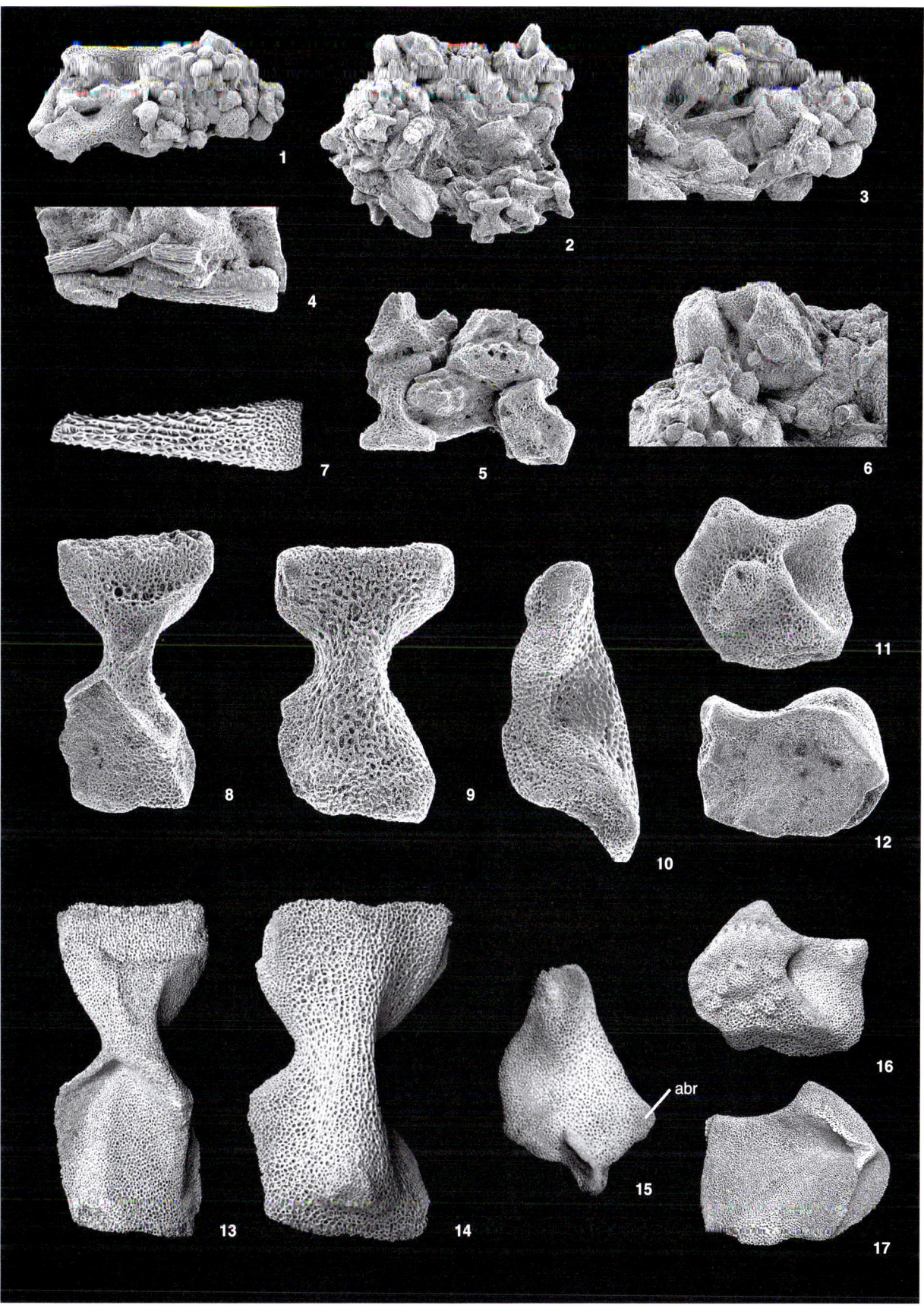

abr

ginal contact) and distribution of spine bases (compare Pl. 20, figs 1–4, 9).

Clade VELATIDA Perrier, 1893
Family PTERASTERIDAE Perrier, 1875

Diagnosis. Velatid asteroids that possess megapaxillae; interradial chevron ossicles elongated, with bifid distal processes; abactinals and distal chevron ossicles strongly imbricate and possess discrete hollows for insertion of connecting muscles.

Genus SAVIGNASTER gen. nov.

Derivation of name. After the village of Savigna, French Jura, where the type material was found.

Diagnosis. A pterasterid that bears megapaxillae with short, stout, vase-shaped pedicels, distally hollow; adambulacrals with short, broad, rod-like lateral extensions, carrying fan-shaped, trellis-like spines at their lateral margin.

Savignaster wardi gen. et sp. nov.
Plate 21; Plate 22, figures 1–3; Plate 23, figures 3–12; Plate 24, figures 1–2, 4, 6–12, 14; Plate 25, figures 1, 4–7, 12–15

Derivation of name. In honour of David Ward, who processed the 1000 kg of clay that yielded all the type material.

Holotype. NHM EE 13610 (Pl. 21, figs 1–2, 5, Pl. 22, figs 1–3).

Paratypes. NHM EE 13611–13632 (Pls 23–25).

Material. Holotype, a partly articulated individual in a nodule with associated abactinals, ambulacrals, adambulacrals and spines (NHM EE 13610); over 300 isolated ossicles, including orals, circumoral, ambulacrals, adambulacrals, terminals and abactinals (all paratypes), from the upper Oxfordian marls (*bifurcatus* Zone, *stenocycloides* Subzone) at Savigna, Jura, France (Hess 1971).

Diagnosis. *Savignaster* that possesses short, very broad adambulacrals that only imbricate slightly.

Description. The holotype is enclosed in a carbonate nodule and is a small, partly dissociated individual that displays adambulacrals, megapaxillae, ambulacrals and abundant spines, some articulated (Pl. 21, figs 1, 2, 5; Pl. 22, figs 1–3). The overall form of the asteroid is not discernible, but by comparison with extant korethrasterids and pterasterids probably possessed five short, stout arms and a broad disc. The adambulacrals (Pl. 22, fig. 3; Pl. 23, figs 2–4, 6, 11, 12) are short and very broad and comprised of a gently curved adradial portion that carries 3–5 spine bases and a long, straight ambulacral extension that bears a single large V-shaped spine base for the actinolateral spine on its termination. The abactinal surface of the adambulacral bears a transverse ridge at the base on which a fused *ada2/3* are present and *ada1a/b*, and the muscle attachments *padam* and *adam* are present on a short distal flange (compare with *Remaster gourdoni*; Pl. 9, fig. 9; Text-fig. 11G). The overall construction of the adambulacrals is similar to that in *Remaster gourdoni* (Pl. 9, figs 9–10) and *Peribolaster biserialis* Fisher, 1905 (Pl. 23, fig. 5). The heads of the ambulacral ossicles (Pl. 24, figs 4, 6) are strongly asymmetrical and possess an enlarged proximal wing, which imbricates proximally over the adjacent ossicle. A short distal flange is also developed, as in *Remaster* Perrier, 1894 and *Pteraster* Müller and Troschel, 1842 (Text-fig. 9). The ambulacral shaft is very narrow and waisted, and the ambulacral base is club-shaped. The area of articulation with the adambulacrals is divided into a flat actinal region separated from the abactinal part of the base by a concave surface, with which *ada2/3* and *ada1a/b* on the adambulacral articulate (see also Pl. 9, fig. 7).

EXPLANATION OF PLATE 20

Fig. 1. *Cheiraster gazellae*, Recent (Table 1), arm section in lateral view, denuded; ×10.

Figs 2–4, 9, 11, 15, 17. *Pontaster tenuispinus*, Recent. Inferomarginal (Fig. 2) in lateral view. Superomarginals (Figs 3, 4) in lateral view. Inferomarginal (Fig. 9) in proximal view. Oral ossicles (Figs 11, 15; ×15) in radial and interradial views, respectively. Circumoral ossicle (Fig. 17; ×15) in abactinal view; figs 2–4, 9, ×20.

Fig. 18. Terminal of *Cheiraster* sp. juv., Recent, Indian Ocean, off Rodrigues (Table 1); ×40.

Figs 5–8, 10, 13, 14, 12, 16, 19. Dissociated ossicles of the benthopectinid *Jurapecten hessi* gen. et sp. nov., upper Oxfordian marls (*bifurcatus* Zone, *stenocycloides* Subzone) at Savigna, near Orgelet, Jura, France. All paratypes.

Figs 5, 8, 12, 16. Inferomarginals in lateral (5; NHM EE 13600; ×25), proximal (8; NHM EE 13603; ×25), actinal (12; NHM EE 13604; ×25) and oblique actinal–lateral (16; NHM EE 13608; ×25) views.

Figs 6–7. Superomarginals in lateral views (6; NHM EE 13601. 7; NHM EE 13602); ×25.

Figs 10, 14. Oral ossicles (NHM EE 13606–13607) in radial (external) and interradial (internal) views, respectively; ×35.

Fig. 13. Circumoral ossicle in abactinal view (NHM EE 13605); ×35.

Fig. 19. Terminal (NHM EE 13609) in abactinal view; ×19.

PLATE 20

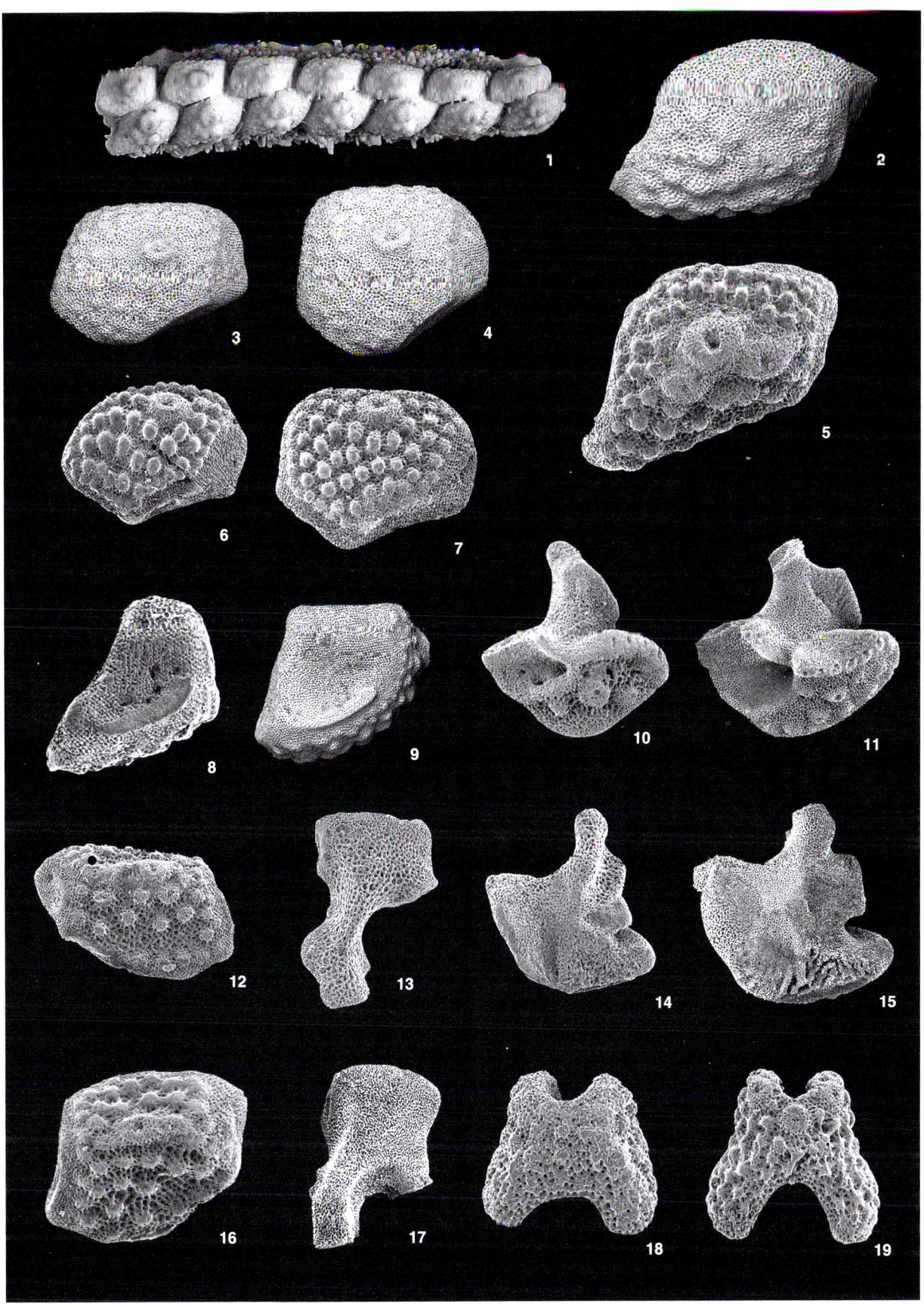

GALE, Fossil and Recent benthopectinids

The oral ossicles (Pl. 23, figs 7–8) comprise an elongated, proximally directed apophyse and a rectangular actinal portion with a short proximal blade. The apophyse (Text-fig. 13D, F) has a deep, long *rvg* and an elongated, large *riom*. Five ridges and grooves form a dentition on the inner (interradial) actinal surface. Four or five small oral spine bases are present on the proximal surface of the blade and a single large, rounded suboral spine base occurs towards the proximal/actinal corner of the radial surface. The shape and construction of the oral of *Savignaster wardi* is close to that of the extant *Pteraster myonotus* Fisher, 1916 (Pl. 23, figs 9–10).

The arrangement of pterasterid interradial ossicles is shown in Text-fig. 35. Two primary interradials are present in the material from Savigna. These (Pl. 25, figs 14–15) have flat actinal surfaces and four elongated processes to contact the primary radials (2, proximally) and abactinal ossicles (2, distally). A median prong-like, narrow process on the distal end of the primary interradial contacts the first (most actinal) pair of chevron plates (Text-fig. 35). The actinal surfaces of the paired processes are notched for muscle insertion, and the notches carry fine labyrinthic stereom (Pl. 25, fig. 13). The primary interradials are similar to those of *Pteraster myonotus* (Pl. 25, figs 10–11). The other abactinal ossicles are megapaxillae, comprising a two-, three- or four-sided basal part and a vase-shaped conical pedicel which is hollow at the tip (Pl. 24, figs 7–12, 14–16; Pl. 25, figs 4–5, 7). The pedicel (Pl. 25, figs 4–5) is ridged along its length and notched near the termination for the attachment of spines. The bases of the abactinals possess 2–4 flattened processes, commonly forming an embayed trapezoidal outline (Pl. 24, figs 7, 10–11). The processes were imbricated in life, and all carry notches for the insertion of interabactinal muscles (e.g. Pl. 24, figs 11, 14; Pl. 25, fig. 7). The trapezoidal bases of these ossicles are very similar to those present in extant *Pteraster myonotus* (Pl. 24, fig. 13). Overall, the abactinal ossicles of *Savignaster wardi* are quite different to those of *Remaster* and *Peribolaster*, which do not possess pedicels and lack any muscularization (Pl. 22, fig. 5; Pl. 25, fig. 2), but close to those of pterasterids. In particular, the vase-shaped pedicels are seen only in the extant genus *Diplopteraster* Verrill, 1880 (Pl. 25, fig. 3).

Interradial chevron ossicles are Y-shaped ossicles that formed the base of the interradial groove in korethrasterids and pterasterids (Text-fig. 35). Those of *S. wardi* possess an elongated and swollen interradial region, which contacted the adjacent chevron plate along the interradius and two asymmetrical processes that contacted abactinal ossicles (Pl. 25, figs 6, 12). These are notched distally for the insertion of muscles. The chevron plates are very similar to those of *Pteraster myonotus* (Pl. 25, figs 8–9), but dis-

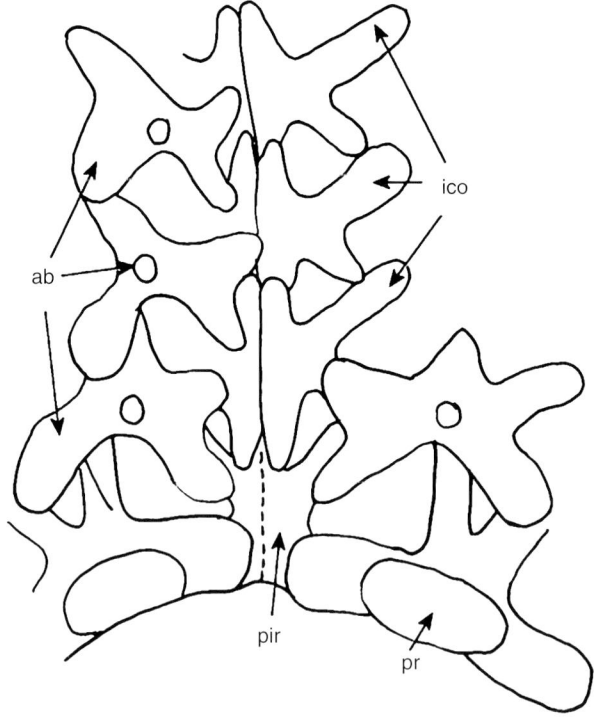

TEXT-FIG. 35. Abactinal interradial structure of *Pteraster obesus myonotus* Fisher, 1916. Recent, Philippines, dissected to expose primary interradial and interradial chevron ossicles (NHM EE13574). Compare with *Savignaster wardi* ossicles in Plate 25, ×10.

similar in shape to those of *Peribolaster* Sladen, 1889 (Pl. 25, fig. 16). Two terminal ossicles present in the Savigna residues probably belong to *Savignaster wardi*. These (Pl. 24, figs 1–2) possess a broad, semicircular termination that carries an array of six distally directed spine pits set on bosses. The divergent proximal processes are delicate and taper proximally; these show some resemblance to those of *Pteraster myonotus* (Pl. 24, fig. 3), and narrow proximal processes on the terminals are characteristic of pterasterids in general (e.g. *Benthaster* Sladen, 1882; see Sladen 1889, pl. 94).

Spines are preserved *in situ* on the holotype. The large actinolateral spines are fan-shaped, concavo-convex and possess an enlarged actinally swollen base (Pl. 21, figs 2, 5). They are made up of 3–5 elongated strips of stereom conjoined by transverse struts, forming a trellis-like structure. The convex side of the

---

EXPLANATION OF PLATE 21

Figs 1–2, 5. *Savignaster wardi* gen. et sp. nov., upper Oxfordian marls (*bifurcatus* Zone, *stenocycloides* Subzone); (horizon 2b) at Savigna, near Orgelet, Jura, France; holotype, NHM EE 13610. Abactinal surface (1), showing vase-shaped abactinal ossicles and abactinal spines, some still in original clusters (C). Actinolateral spines (2, 5), close to original position on margin, attached to adambulacral extensions. Fig. 1, ×35; Fig. 2, ×55.

Figs 3–4. *Remaster gourdoni*, Recent (Table 1), actinolateral spines in abactinal and actinal views, respectively, for comparison. Figs 3–4, ×60.

PLATE 21

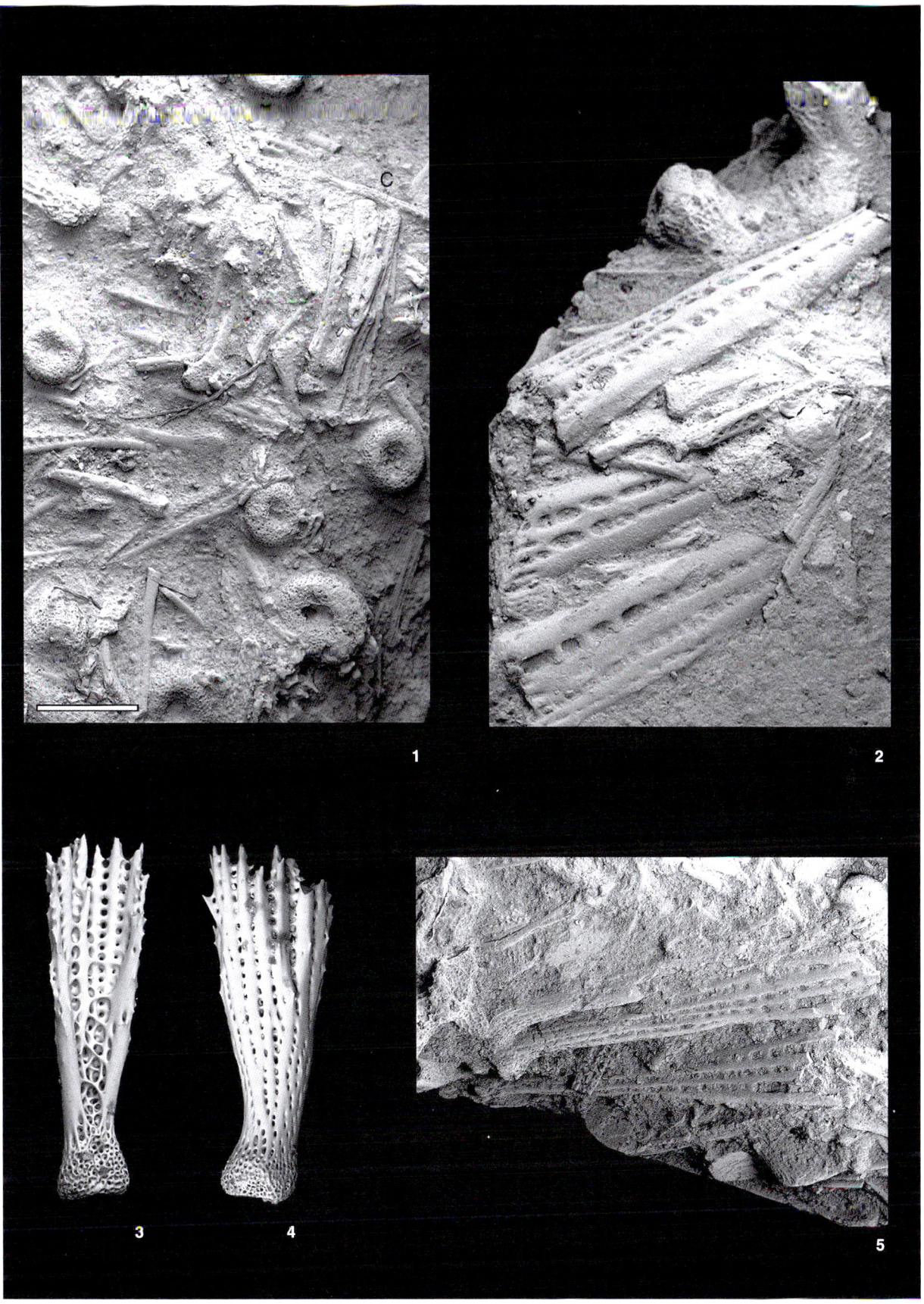

GALE, Fossil pterasterid, Recent korethrasterids

spine is directed actinally. These actinolateral spines are very similar to those present in *Remaster gourdoni* (Pl. 21, figs 3–4; Pl. 22, fig. 6). Spines from the base of the adambulacrals are cylindrical, with longitudinal grooves and rows of pits (Pl. 22, fig. 1). Spines of the abactinal ossicles are preserved as clusters close to their original attachment sites, and each abactinal carried 5–10 spines (Pl. 21, fig. 1). These are elongated and narrow, with an enlarged base and a ladder-like blade with transverse struts of stereom. They are very similar to those of extant *Remaster* (Pl. 22, fig. 5) and were probably united by a membrane as in extant *Remaster* (e.g. Fell and Clark 1959, figs 1–2) and *Peribolaster* (Clark 1963, fig. 15A).

Remarks. The presence of specialized, distally bifid interradial chevron ossicles, abactinal megapaxillae with vase-shaped pedicels and interabactinal muscle facets on the imbricating surfaces of the distal chevron plates and abactinals indicate that *Savignaster wardi* belongs to the Pterasteridae. However, the form of the adambulacrals with elongated lateral extensions, the absence of large primary radials with laterally expanded heads and the fan-like, trellised form of the lateral adambulacral spines are closely similar to those developed in the korethrasterid genera *Remaster* and *Peribolaster*. *Savignaster wardi* is therefore the most basal pterasterid known and is positioned as a phylogenetic intermediate between *Peribolaster* and the Pterasteridae.

From the above description, it is possible to infer that *Savignaster wardi* lacked the abactinal canopy found in other pterasterids, because the enlarged, highly muscularized form of the abactinal and lateral adambulacral spines effectively precludes the development of an enclosing membrane. However, the presence of muscles between the abactinal ossicles indicates that *Savignaster wardi* was capable of contracting the upper surface of the body in a manner similar to that of the living *Pteraster tesselatus* (see Nance and Braithwaite 1972). This action reduces coelomic volume and causes the papulae to balloon out, significantly increasing the respiratory surface. The presence of this muscularization in *Savignaster wardi*, a pterasterid that lacked an abactinal canopy, permits the suggestion that improved oxygen uptake was an important selective drive behind evolution of the highly specialized pterasterids.

Savignaster trimbachensis gen. et sp. nov.
Plate 23, figures 1–2

Derivation of name. From the locality of Trimbach in Switzerland.

Holotype. NMB M 11143 (Pl. 23, figs 1–2). This adambulacral is from the upper Oxfordian (Birmenstorf Member) at Rumperl, south-west of Trimbach, near Olten, Canton Solothurn, Switzerland.

Diagnosis. *Savignaster* with obliquely flattened, strongly imbricating adambulacrals, which possess a strong articular structure at the base of the adambulacral extension.

Description. The adambulacral ossicle (Pl. 23, figs 1–2) is obliquely flattened and J-shaped and possesses a stout, straight-sided adambulacral extension, on the termination of which is an attachment site for the actinolateral spine. On the distal surface of the extension a large, distal *adexa* (special articulation surface between adambulacral extensions; see Text-fig. 12) is developed. The distal site of insertion of the adadm is concave. On the distal surface of the outer face five large, pustule-like spine bases are present. The flattened form of the ossicle indicates that the adambulacrals imbricated strongly proximally. A single large broken adambulacral head and shaft amongst the Trimbach material probably belongs to this species. This has a large distal flange, strongly developed dentition and a large insertion site for the *itam*.

Remarks. *Savignaster trimbachensis* differs from *Savignaster wardi* in the flattened, strongly imbricating form of the adambulacrals and in the possession of a large articulation surface between the adambulacral extensions.

Order FORCIPULATIDA Perrier, 1884

Family TERMINASTERIDAE fam. nov.

Diagnosis. Forcipulatid asteroids in which the arm is made up of only seven ossicle rows (one radial, two adradial, two superomarginal and two inferomarginal) all of which alternate. Superomarginal and abactinal ossicles of the arm all possess six articulation surfaces, the infero-

---

EXPLANATION OF PLATE 22

Figs 1–3. *Savignaster wardi* gen. et sp. nov., upper Oxfordian marls (*bifurcatus* Zone, *stenocycloides* Subzone) Zone (Horizon 2b), Savigna, near Orgelet, Jura, France; holotype, NHM EE 13610. Adambulacral spines (compare Pl. 21), disarticulated ambulacrals (for comparison with Pl. 24) and adambulacral, respectively. Figs 1, 3 × 55; Fig. 2 × 40.

Fig. 4. *Diplopteraster verrucosus*, Recent, Argentina (Table 1), denuded abactinal surface, ×5.

Figs 5–6. *Remaster gourdoni*, Recent, Falklands, partly denuded abactinal surface (for comparison with Pl. 21, fig. 1) and actinal interradial view, showing adambulacrals, chevron ossicles and actinolateral spines, respectively; Fig. 5 × 65; Fig. 6 × 25.

PLATE 22

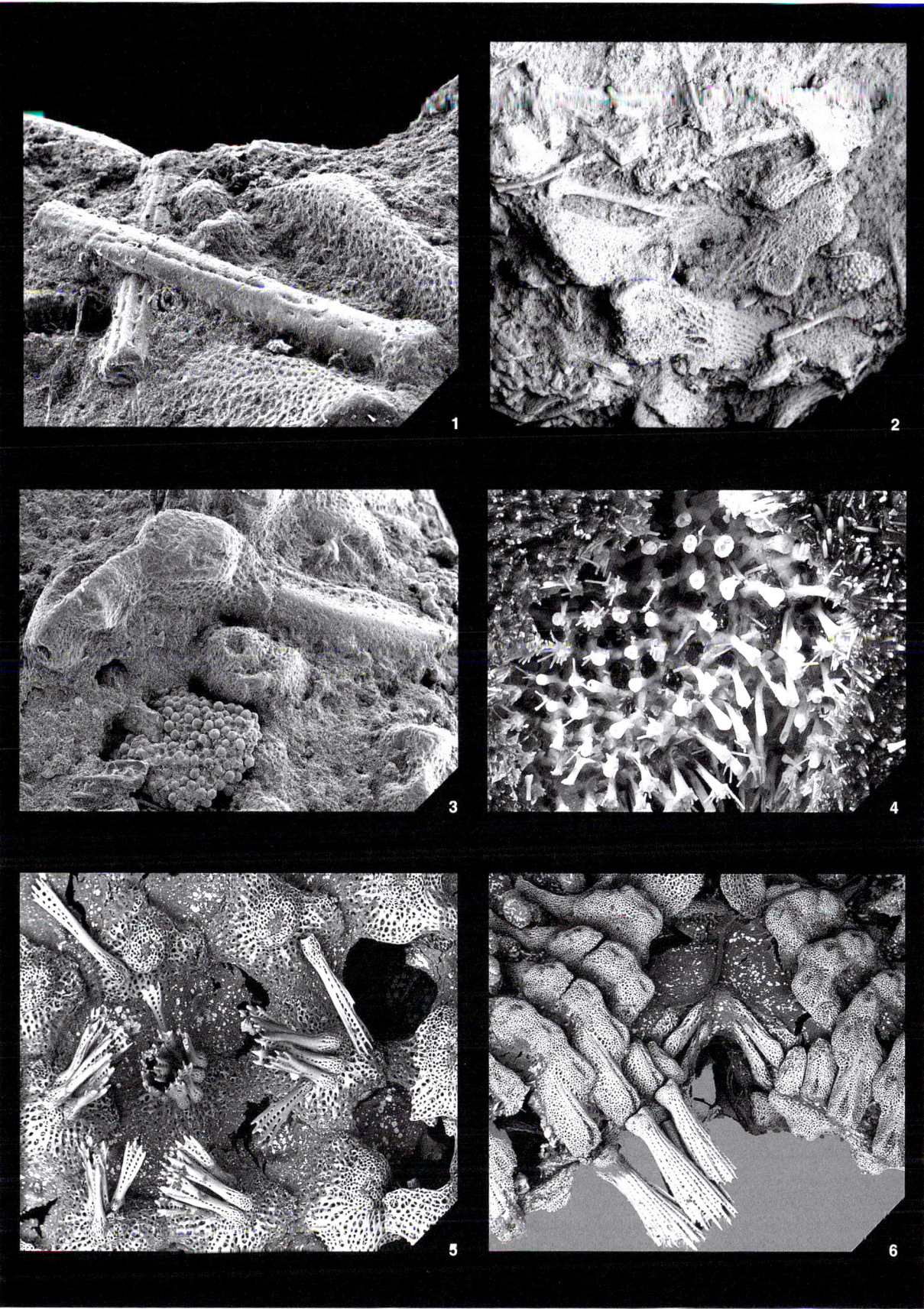

GALE, Fossil and Recent pterasterids and korethrasterids

TABLE 8. Identification of material included in *Terminaster cancriformis* by Hess (1974).

Hess (1974)	Interpretation of Hess	Identification herein
Pl. 1, fig. 9	Oral of *Terminaster*	Benthopectinid oral
Pl. 2, fig. 1	Adamb of *Terminaster*	Adamb of *Plumaster*
Pl. 2, fig. 5	Superomarginal of *Terminaster*	Benthopectinid marginal, *Jurapecten hessi*
Pl. 3, fig. 7	Centrodorsal ? of *Terminaster*	Abactinal of *Plumaster*

marginals five. Surface sculpture of marginals and abactinals consists of dense rugosities conjoined by struts of stereom.

Genus TERMINASTER Hess, 1971

Terminaster cancriformis (Quenstedt, 1876)
Plate 26, figures 1–3, 5–6, 9, 11–12; Plate 27, figures 1–3, 6–9, 11–16

> 1876 *Asterias cancriformis* Quenstedt, p. 49, pl. 91, fig. 155.
> p. 1971 *Terminaster cancriformis* (Quenstedt); Hess, pp. 647–659 (*pars*), text-figs 1–5; pls 1–3, non pl. 1, fig. 9; pl. 2, figs 1, 5; pl. 3, fig. 7.
> 2009 *Terminaster cancriformis* (Quenstedt); Villier *et al.*, p. 390, figs 3–4.

Description. The species is redescribed on the basis of extensive new material from the upper Oxfordian of Savigna, Fench Jura (Hess 1971), which includes clusters of associated and partly articulated ossicles, some with spines and over 600 isolated ossicles, including abactinals, marginals, adambulacrals, ambulacrals, orals, circumorals and terminals. This material permits a reassessment of the ossicles assigned to *Terminaster cancriformis* by Hess (1971), not all of which belong to this species (Table 8). The arms are elongated and narrow, the disc small (Hess 1971; Villier *et al.* 2009) and the tube feet uniserial. The abactinal surface of the disc is made up of five primary interradials, five primary radials and one centrale. The centrale is oval, with a radiating sculpture of irregular rugosities and short articular processes on the margin (Pl. 27, fig. 14). The primary interradials are rectangular in outline with a single large, abactinally positioned spine base and are inset beneath the primary radials (Pl. 27, fig. 11). The primary radials (Hess 1971, pl. 3, fig. 4) are trapezoidal in outline and lack large spine attachment sites.

The arm is made up of seven rows of plates, comprising two inferomarginal rows, two superomarginal rows, two adradial rows and one radial row (Hess 1971; Text-fig. 36A herein). The individual figured by Villier *et al.* (2009) has only small adradials and is probably immature. The plates of each type are morphologically distinct and can be readily identified amongst the isolated asteroid debris in screened samples. The inferomarginals (Pl. 27, figs 6, 15–16) imbricate weakly proximally and possess a single central large spine base that has discrete proximal and distal crescentic articular surfaces. The inferomarginals possess two abactinal processes that are imbricated by corresponding structures on the superomarginals. The adradials (Pl. 27, fig. 12) are smaller than the radials and superomarginals, and their outer surfaces display four short articular processes that in life articulate with the underside of the corresponding processes on the radials and superomarginals. The adradials do not carry an enlarged spine base, but are covered with relatively fine rugosities. The radials (Pl. 27, figs 7–9) are oval in outline, imbricate proximally, and larger ossicles possess short, paired distal and proximal processes, which articulate with the adjacent adradials. Each radial carries a single large circular spine base, positioned on the perradial line slightly proximal to the central position. All abactinal and marginal ossicles possess a distinctive sculpture of small, raised

EXPLANATION OF PLATE 23

Figs 1–2. *Savignaster trimbachensis* gen. et sp. nov., holotype (NMB M 11143), upper Oxfordian, Birmenstorf Member, Rumperl, south-west of Trimbach, near Olten, Canton Solothurn, Switzerland, ambulacral ossicle, in actinal and abactinal views, respectively.

Figs 3–4, 6–9, 11–12. *Savignaster wardi* gen. et sp. nov., upper Oxfordian marls (*bifurcatus* Zone, *stenocycloides* Subzone) at Savigna, near Orgelet, Jura, France, from horizons 1 (Figs 11–12) and 2b (Figs 3–4, 6–8), all paratypes.

Figs 3–4. Adambulacral base (broken, NHM EE 13611) in abactinal and actinal views, respectively.

Fig. 6. Small adambulacral (NHM EE 13612) in actinal aspect.

Figs 7–8. Oral ossicles (NHM EE 13613–13614) in interradial and radial aspects, respectively.

Figs 11–12. Ambulacral ossicle (NHM EE 13615) in abactinal and actinal views, respectively.

Fig. 5. *Peribolaster biserialis*, Recent, Bering Sea (Table 1), adambulacral in actinal view.

Figs 9–10. Oral ossicle of *Pteraster obesus myonotus* Fisher, Recent, Philippines (Table 1) in radial and interradial views, respectively.

Figs 1–2, ×13; Figs 5–6, ×35; Figs 3–4, ×50; Figs 7–8, ×17; Figs 9–10, ×20; Figs 11–12, ×25.

PLATE 23

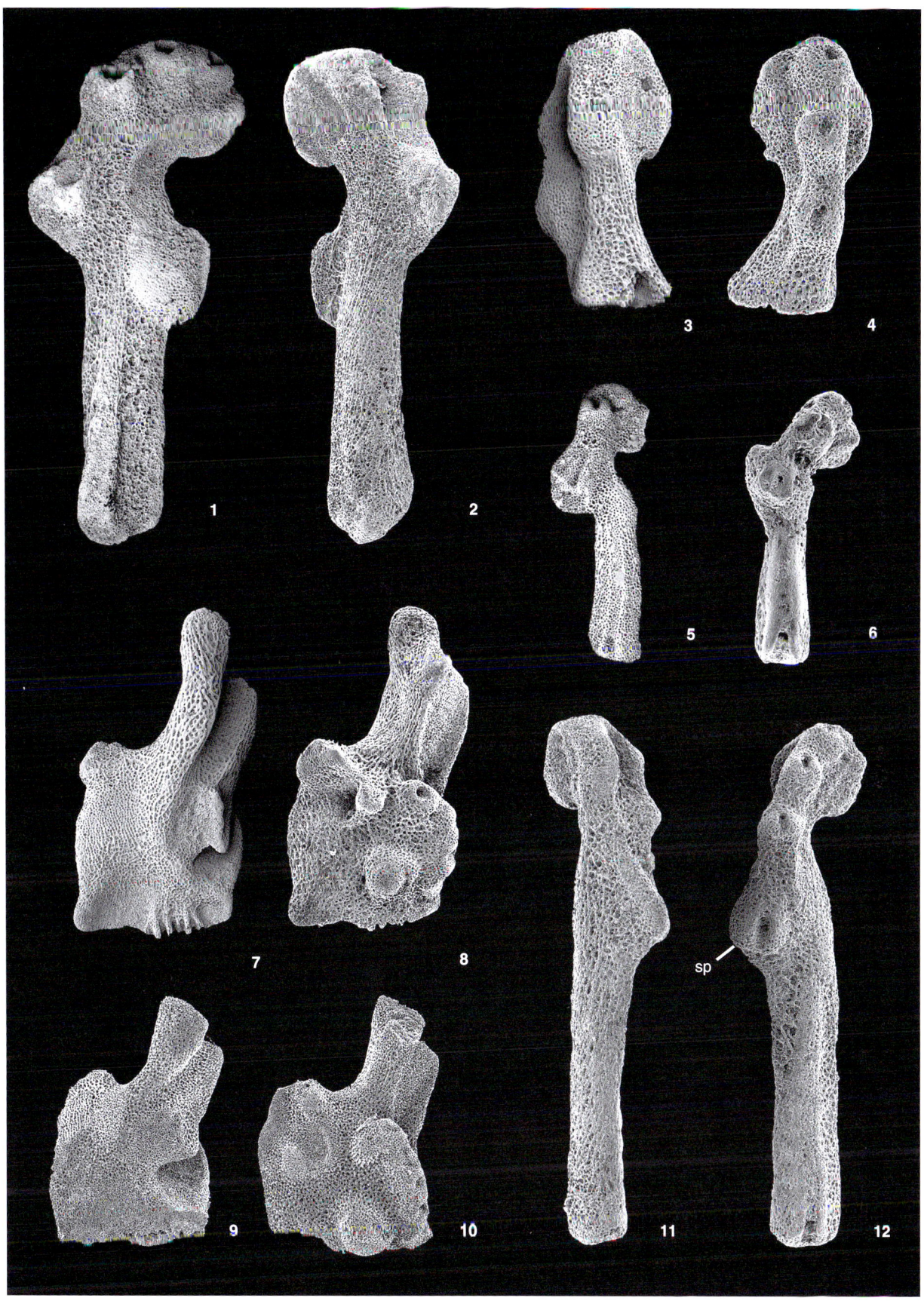

GALE, Fossil and Recent pterasterids

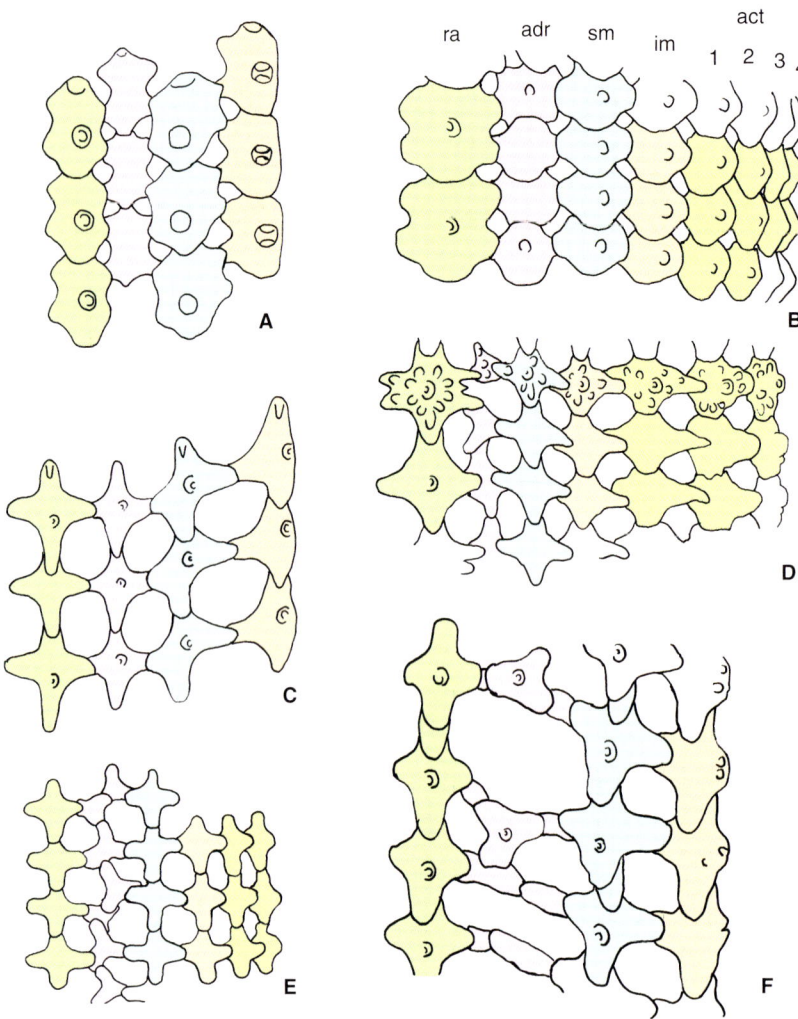

ra adr sm im act
              1 2 3 4

TEXT-FIG. 36. Abactinal and marginal plating in forcipulatids; terminasterid (A); zoroasterids (B, D); pedicellasterid (E); asteriids (C, F), homologies colour coded. A, *Terminaster cancriformis*, reconstruction of plating of arm. B, *Zoroaster ophiurus* Fisher, 1928 (after Fisher 1928, pl. 13, fig. 2). C, undescribed asteriid, Maastrichtian of Morocco. D, *Myxoderma platyacanthum* Fisher, 1928 (after Fisher 1928, pl. 16, fig 2a). E, *Pedicellaster magister* Fisher, 1923 (after Fisher 1928, pl. 28, fig. 1b). F, *Stylasterias forreri* (de Loriol, 1887) (juvenile, after Fisher 1928, pl. 44, fig. 2). The arrangement and imbrication pattern of arm ossicles is quite consistent in the forcipulatid clade; the adradial ossicles are *inset* between radials and superomarginals, which in turn imbricate *over* the inferomarginals. The first actinal row is imbricated by the inferomarginals, and successive actinal rows are imbricated by more adradial ones. This is here called the Forcipulatid Plating Rule. A, C, ×20; all other, ×10.

bosses conjoined by narrow ridges of stereom. In life, the bosses probably bore small spines. The arrangement of the arm ossicles (Text-fig. 35A) can be reconstructed with reference to the individuals figured by Hess (1971).

The adambulacrals are all similar in shape and rectangular in outline from an actinal aspect (Pl. 26, fig. 5). The actinal surface bears a centrally placed, very large, bifid spine base and, adradially, two smaller spine bases. The ambulacrals (Pl. 26, figs

---

EXPLANATION OF PLATE 24

Figs 1–2, 4, 6–12, 14–16. *Savignaster wardi* gen. et sp. nov., upper Oxfordian marls (*bifurcatus* Zone, *stenocycloides* Subzone) (horizon 2b), Savigna, near Orgelet, Jura, France. All paratypes.

Figs 1–2. Terminal ossicle (NHM EE 13616) in abactinal and actinal views, respectively.

Figs 4, 6. Ambulacral ossicles (NHM EE 13617–13618) in abactinal and actinal views, respectively.

Figs 7–12, 14–16. Abactinal ossicles in abactinal aspect (NHM EE 13619–13626). Specimens in 7, 10–11 and 14 are from a radial position (embayed trapezoidal outline, symmetrical); the others from interradial positions. Note sites for insertion of interabactinal muscles (*m*).

Fig. 5. *Peribolaster biserialis*, Recent, Bering Sea. Ambulacral in actinal view.

Figs 3, 13. *Pteraster obesus myonotus*, Recent, Philippines (Table 1), terminal ossicle in abactinal view and abactinal ossicle, close to radial line, in abactinal aspect, respectively.

Figs 1–2, ×20; Fig. 3, ×40; Fig. 4, ×35; Figs 5–12, 14–16, ×30; Fig. 13, ×25.

PLATE 24

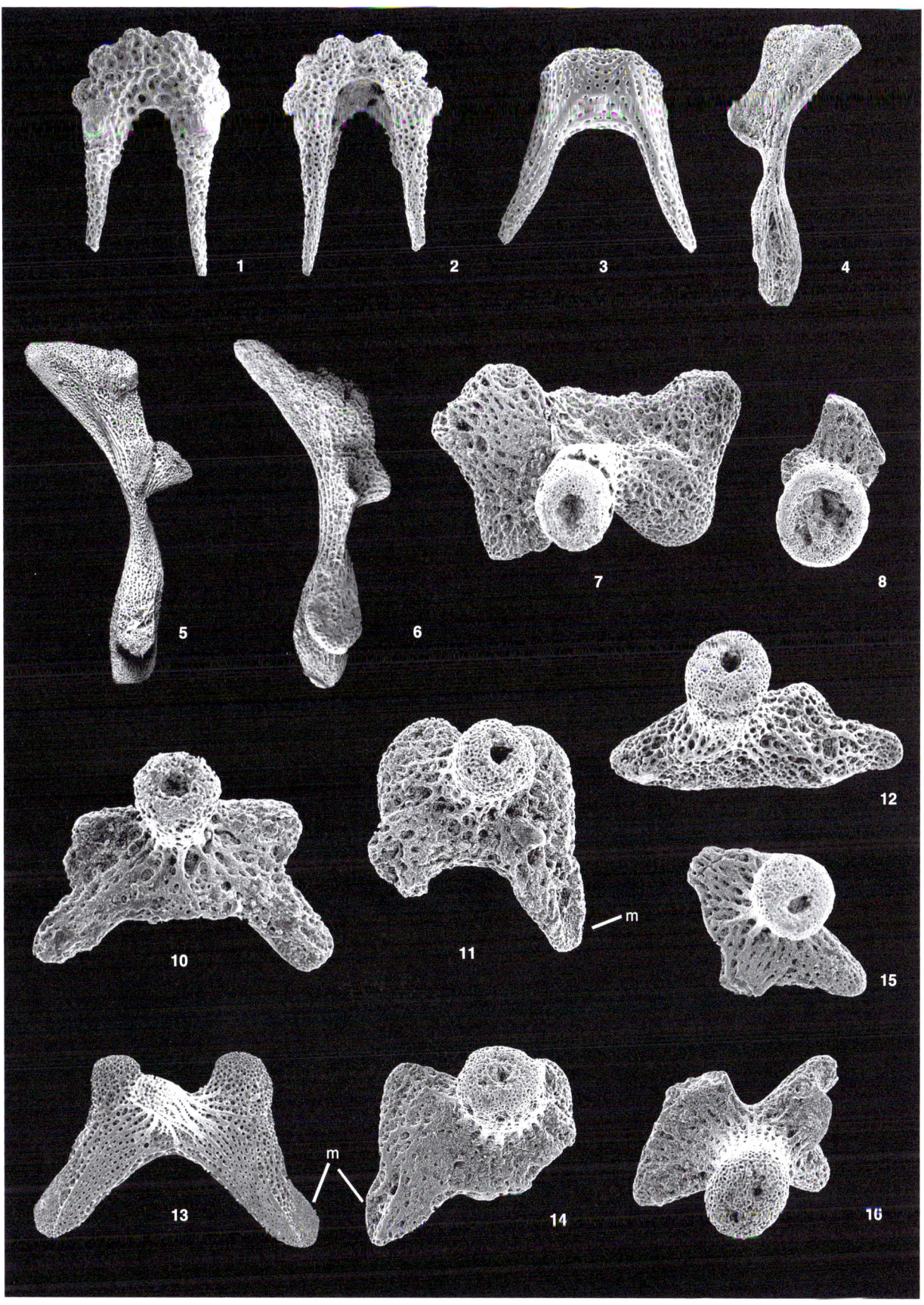

GALE, Fossil and Recent pterasterids and korethrasterids

11–12; see also Hess 1971, pl. 1, fig. 10) are moderately elongated, and the axis is angled slightly relative to the adradial contact with the adjacent ambulacral row. The ridge for adambulacral contact is strongly concave, and short wings for padam and dadam insertion are present. The oral ossicles (Pl. 26, figs 6, 9) have a tall apophyse, and the *rvg* and *nrg* are oblique to the interradial contact. An elongated articulation/muscle insertion surface for the first adambulacral dominates the distal part of the outer (radial) face, and two attachment sites for oral spines are present on the proximal blade. Two discrete regions of dentition on the interradial face are separated by an insertion site for an actinal interoral muscle. The oral ossicle is very similar to that of *Zoroaster fulgens* (Pl. 26, figs 7, 10) differing principally in the higher apophyse and minor differences of shape. Circumorals probably belonging to *Terminaster cancriformis* (Pl. 26, fig. 3) possess the characteristically short *pcp* of forcipulatids, and the relative height of the two processes matches that of the corresponding structures on the oral ossicles. The terminal ossicles are highly distinctive, as they are very large and elongated and carry large spine bases along the actinal margins and, less commonly, on the abactinal surface (Pl. 27, figs 1–2). The actinal groove is very deep and sometimes contains remains of distal adambulacrals and ambulacrals (Hess 1971; Pl. 27, fig. 3 herein). The spines of *Terminaster cancriformis* (Pl. 26, figs 1–2, 5) are conical and taper rapidly; they bear undulating length-parallel ridges of stereom. They are similar to marginal and abactinal spines of *Myxoderma sacculatum ectenes* Fisher, 1919 (Pl. 26, fig. 8).

Remarks. *Terminaster* was assigned to the Zoroasteridae by Hess (1971) and placed centrally within the family at base of the 'imbricate' zoroasterids by Mah (2007). The genus was considered to be either a basal forcipulatid, or paedomorphically derived from the Zoroasteridae by Villier *et al.* (2009). The new material permits detailed comparison of ossicle morphology to be made with other forcipulatids and thus a re-evaluation of the affinities and taxonomic position of *Terminaster*.

In forcipulatid asteroids, the arrangement and pattern of imbrication of the abactinal, marginal and actinal ossicles in the arm is quite consistent and can be called the Forcipulatid Plating Rule (Text-fig. 36). First, the radial and superomarginal ossicles imbricate over the adradial plate row (or rows), which are inset and often do not carry large spines (see Fisher 1928, pls 13, 15, 44; Text-fig. 36 herein). Second, the superomarginal ossicles always imbricate over the inferomarginals; it is thus always possible to identify radial and superomarginal ossicles because only these overlie both adjacent ad- and abradial plate rows. Fisher (1928) used this arrangement to identify plate rows in forcipulatid asteroids. The inferomarginals are imbricated by the superomarginals and themselves imbricate the most adradial actinal row; successive actinal rows imbricate abradially (Text-fig. 36B, D, E). The origin of the imbrication pattern probably reflects the order of appearance of ossicles during early postmetamorphosis development of the skeleton, with radials and superomarginals appearing first and adradials and inferomarginals added subsequently (Kano *et al.* 1974). *Terminaster cancriformis* has typical forcipulatid ossicle arrangement and follows the plating rule. The plating rule can also be used to identify other fossil forcipulatids. Thus, the supposed echinasterid *Protothyraster priscus* (de Loriol, 1873) (Hess, 1970, fig. 15) can be placed within the Forcipulatida because it has an inset row of adradial ossicles.

Addition of extra rows of ossicles tends to occur either in the inset adradial zone (Asteriidae, Pedicellasteridae) or in the actinal zone (Zoroasteridae, Pedicellasteridae). The inferomarginals are identifiable because they are usually

---

EXPLANATION OF PLATE 25

Figs 1, 4–7, 12–15. *Savignaster wardi* gen et sp. nov., upper Oxfordian marls (*bifurcatus* Zone, *stenocycloides* Subzone) Savigna, near Orgelet, Jura, France, from horizons 1 (Fig. 6), 2a (Figs 4–5, 12) and 2b (Figs 13–15). All paratypes.

Figs 1, 6, 12. Interradial chevron ossicles (NHM EE 13627–13628) (see Text-fig. 35) in abactinal view. Fig. 1 is an enlargement of a muscle insertion site of ossicle in Fig. 12.

Figs 4–5, 7. Abactinal ossicles (NHM EE 13629–13631) in lateral (4–5) and actinal (7) views. Note muscle insertion site (*m*).

Figs 13–15. Primary interradial ossicle (NHM EE 13632) in abactinal and actinal views, respectively; Figure 13 is an enlargement of the muscle insertion site with the primary radial ossicle (see Text-fig. 35).

Figs 2, 16–19. *Peribolaster biserialis*, Recent, Bering Sea (Table 1). Abactinal ossicle (2) in abactinal view. Interradial chevron ossicle (16) in abactinal view. Primary interradial ossicle (17–19) in abactinal and actinal views, respectively. 17, enlargement of distal process for attachment with primary radial ossicle. Note the absence of muscle insertion site.

Fig. 3. Abactinal ossicle of *Diplopteraster verrucosus* (Table 1) in lateral aspect.

Figs 8–11. *Pteraster obesus myonotus*, Recent, Philippines, abactinal interradial chevron ossicle in abactinal aspect. 10–11, primary interradial ossicle in abactinal and actinal aspects, respectively. 9, enlargement of muscle insertion site on process for articulation with primary radial ossicle.

Figs 10–11, ×8; Figs 3, 8, 14–15, ×20; Figs 4–5, ×30; Figs 2, 6–7, ×35; Figs 12, 16, ×40; Fig. 9, ×60; Fig. 1, ×100; Fig. 17, ×120.

PLATE 25

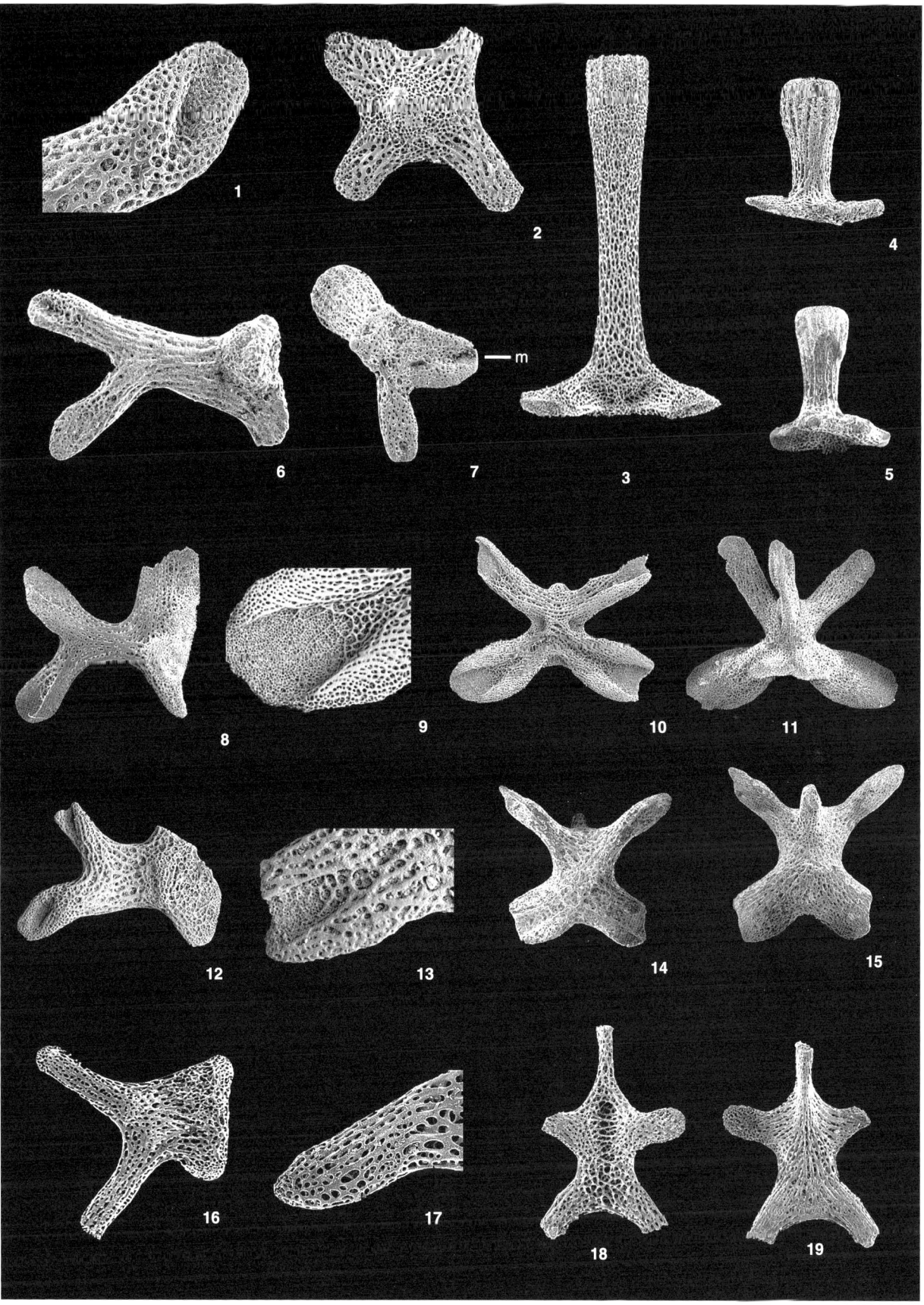

GALE, Fossil and Recent pterasterids and korethrasterids

large, spine-bearing ossicles adjacent to and imbricated by the superomarginals. However, in zoroasterids and some pedicellasterids, inferomarginals are morphologically identical to and only slightly larger than the actinals (Text-fig. 36B, D, E). The argument that all zoroasterids possess only one row of marginals (Blake 1987; Blake and Elliot 2003; Mah 2007) is based upon the observation that young, postmetamophosis *Zoroaster* possess only a single row of marginals and that the row of ossicles identified as inferomarginals by Fisher and others (e.g. Sumida *et al.* 2001) do not reach the terminal ossicle, supposedly a defining feature of marginal ossicles (Blake 1988*a, b*; Mooi and David 2000). However, the Forcipulatid Plating Rule clearly applies to zoroasterids and allows identification of the row of ossicles imbricated by the superomarginals as inferomarginals (Text-fig. 36B, D). Superomarginals develop first in zoroasterids, and inferomarginal ossicles are only added at a later growth stage, forming initially between the superomarginals and adambulacrals in the mid-arm (Sumida *et al.* 2001, fig. 11I). The definition of marginal ossicles as a continuous row in contact with the terminal ossicle is perhaps too rigid.

The morphology of the circumoral ossicles, the presence of an adoral carina and an actinostome (Hess 1971; Blake 1987; Villier *et al.* 2009) are all forcipulatid features found in *Terminaster cancriformis*. Additionally, that species shares a number of characters with the zoroasterids.

These include:

– the shape and detailed structure of the oral ossicles;
– overall body form, with long arms and a small disc (although this feature is developed in the forcipulatids more generally);
– only a single row of adradials present, lacking a large central spine base;
– elongated terminal ossicles.

A large number of features of *Terminaster cancriformis* are apparently paedomorphic and possibly derived from an ancestor by accelerated maturation (McKinney and McNamara 1991):

– Very large terminal ossicles are characteristic of juvenile asteroids in general, because these ossicles appear very early in ontogeny. Additionally, juvenile *Myxoderma* Fisher, 1905 and *Zoroaster* possess huge terminals (Fisher 1928, pl. 25; Blake and Elliot 2003, fig. 1: 1–4).
– Juvenile *Zoroaster fulgens* have adambulacrals that are all similar (e.g. Blake and Elliot 2003, fig. 1: 3) like those of *Terminaster*; the characteristically alternately carinate and noncarinate plates are an adult feature in zoroasterids.
– The small number of plate rows in the arm (7) is also a juvenile character, as the number of plate rows in forcipulatid arms increases with ontogeny.
– The retention of large primary abactinal ossicles, lacking small intervening plates, is also a juvenile feature of zoroasterids (Sumida *et al.* 2001; Blake and Elliot 2003).
– Juvenile zoroasterids initially possess a single large adambulacral spine on each ossicle (Sumida *et al.* 2001, fig. 11I), the condition seen in *Terminaster cancriformis*.
– Juvenile zoroasterids and asteriids possess proportionately elongated ambulacrals; the short, compressed arrangement seen in adults develops later in ontogeny.
– Juvenile zoroasterids have biserial tube feet; they only become quadriserial in the adult.

Other characteristics of *Terminaster* are more likely to be peramorphic:

– The double articulations between all arm ossicles are not developed in any other known forcipulatid; only radial ossicles in *Zoroaster* and *Myxoderma* possess six articular surfaces (e.g. Fisher 1928, pl. 13, figs 1–3).
– The pervasive alternation of the abactinals and the marginals is unusual or unique amongst forcipulatids,

---

EXPLANATION OF PLATE 26

Figs 1–3, 5–6, 9. *Terminaster cancriformis*, upper Oxfordian marls (*bifurcatus* Zone, *stenocycloides* Subzone) at Savigna, near Orgelet, Jura, France; group of associated arm ossicles and spines, partially articulated (NHM EE 13633; Figs 1–2, 5), including superomarginals and ambulacrals (1), adambulacrals (5) and marginal spines. 3, circumoral ossicle in abactinal view (NHM EE 13643). 6, 9, oral ossicle (NHM EE 13635) in radial and interradial aspects, respectively.

Figs 11–12. Ambulacral ossicle of *T. cancriformis* from Terres Rouges, Switzerland, the original of Hess (1971, pl. 1, fig. 10), (NHM B M 9036), in abactinal and actinal aspects, respectively.

Figs 4, 7. 10. *Zoroaster fulgens*, Recent, north-east Atlantic. 4, circumoral ossicle in abactinal aspect. 7, 10, oral ossicle in radial and interradial views, respectively.

Fig. 8. *Myxoderma sacculatum ectenes* Fisher, 1919*b*, north-east Pacific (Table 1), superomarginal spine (Fig. 8).

Figs 1, 6, 9, ×25; Fig. 2, ×40; Fig. 5, ×65; Fig. 3, ×35; Figs 4, 7, 10, ×15; Figs 11–12, ×40; Fig. 8, ×20.

PLATE 26

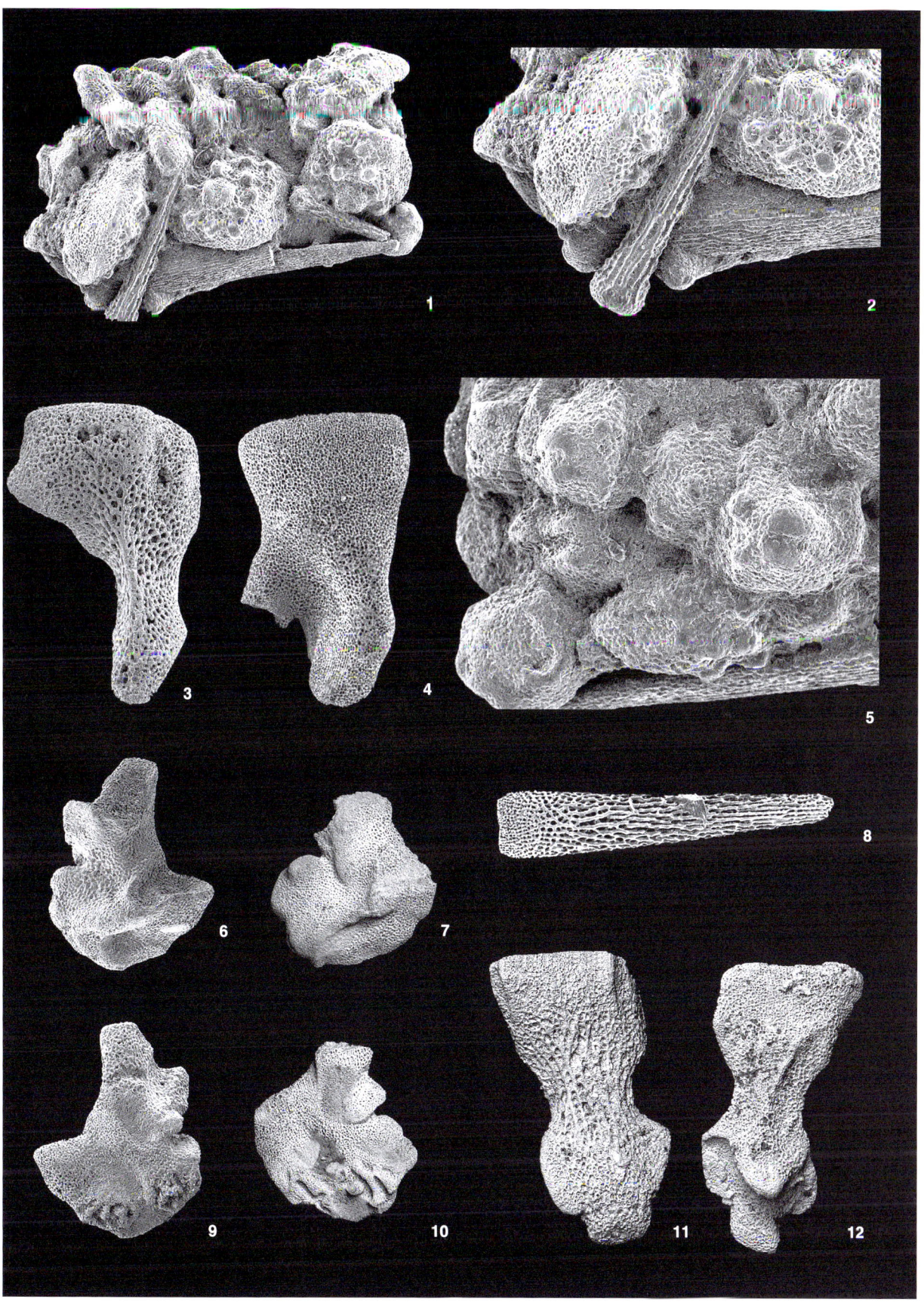

GALE, Fossil terminasterid and Recent zoroasterids

in which plate rows tend to be transverse and nonimbricating.

– The highly distinctive reticulate sculpture of the marginals and abactinal ossicles comprises small spine bases and conjoining ridges.

As pointed out by Villier *et al.* (2009), *Terminaster* has two most likely relationships with zoroasterids; it either represents a basal form (identified by Villier *et al.* 2009 by cladistic analysis as a potentially basal forcipulatid) or it represents a paedomorphic clade derived from within the Zoroasteridae or a related family. Without detailed knowledge of the ontogeny of all taxa considered, cladistic analysis can provide only limited resolution of this type of problem (see Kitching *et al.* 1999). However, the detailed ossicle morphology presented here provides useful evidence with which to test the two relationships. In particular, the very close similarities in the morphology of the oral ossicles (Pl. 26, figs 6–7, 9–10) between *Zoroaster*, *Myxoderma* and *Terminaster* include a number of zoroasterid synapomorphies (presence of dentition, symmetrically placed on either side of an actinal interoral muscle; nerve ring groove and ring vessel groove strongly oblique to interradial face). This tends to support the hypothesis that *Terminaster* was paedomorphically derived from a zoroasterid during or prior to the Middle Jurassic.

INCERTAE SEDIS

Family TRICHASTEROPSIDAE Blake and Hagdorn, 2003

Diagnosis. Adambulacrals broad, short, broadening distally towards arm tip, bearing single transverse row of spine bases; single marginal row; short adoral carina present.

Constituent genera. *Trichasteropsis* and *Berckhemeraster*.

Family MIGMASTERIDAE fam. nov.

Diagnosis. Short, rapidly tapering arms; blocky paired supero- and inferomarginals; a single large interradial inferomarginal which is inset from the ambitus; oval, transversely broad adambulacrals that carry four or five transversely arranged, blade-like spines adradially; abradially, they each bear a single large, blade-like spine on the distal margin.

Constituent genus. *Migmaster*.

Family XYLOPLACIDAE Baker, Rowe and Clark, 1986

Genus ANKYLOPLAX gen. nov.

Derivation of name. From the Greek *ankylos*, meaning fused, with reference to the oral/circumoral ossicle and the adambulacrals.

Type species. *Xyloplax janetae* Mah, 2006.

Diagnosis. Xyloplacidae in which the oral and circumoral ossicles are fused along the length of the oral apophysis. Single elongated, scythe blade-shaped adambulacral present in each half radius, carrying 6–8 spines along its ambital edge. Only three ambulacrals present, which do not articulate with the adambulacrals, but contact the inner surface of the abactinals (Text-fig. 33A).

EXPLANATION OF PLATE 27

Figs 1–3, 6–9, 11–16. *Terminaster cancriformis*, upper Oxfordian marls (*bifurcatus* Zone, *stenocycloides* Subzone) at Savigna, near Orgelet, Jura, France.

Figs 1–3. Terminal ossicle (NHM EE 13636) in actinal, abactinal and enlarged actinal views, respectively, to show small ambulacrals and adambulacrals of arm tip *in situ* within groove on terminal ossicle.

Figs 6, 15–16. Inferomarginal ossicles (NHM EE 13637, EE 13645 and EE 13646, respectively), in external lateral (6, 16) and internal (15) views.

Figs 7–9. Radial ossicles (NHM EE 13638–13640) in abactinal (7, 9) and internal (8) aspects.

Fig. 11. Primary interradial ossicle (NHM EE 13641) in abactinal view.

Fig. 12. Adradial ossicle (NHM EE 13642) in abactinal aspect.

Fig. 13. Superomarginal ossicle (NHM EE 13643) in lateral/abactinal aspect.

Fig. 14. Centrale (NHM EE 13644) in abactinal view.

Figs 4, 10. *Myxoderma sacculatum ectenes*, Recent, north-east Pacific. Radial ossicle (4) in abactinal view. Primary interradial ossicle (11) in abactinal view.

Fig. 5. *Zoroaster fulgens*, north-east Atlantic, superomarginal ossicle.

Fig. 12, ×10; Figs 10, 13, ×15; Figs 1–2, 4–5, 8, 14–15, ×20; Figs 6–7, 9, 16, ×25; Fig. 11, ×35; Fig. 3, ×65.

PLATE 27

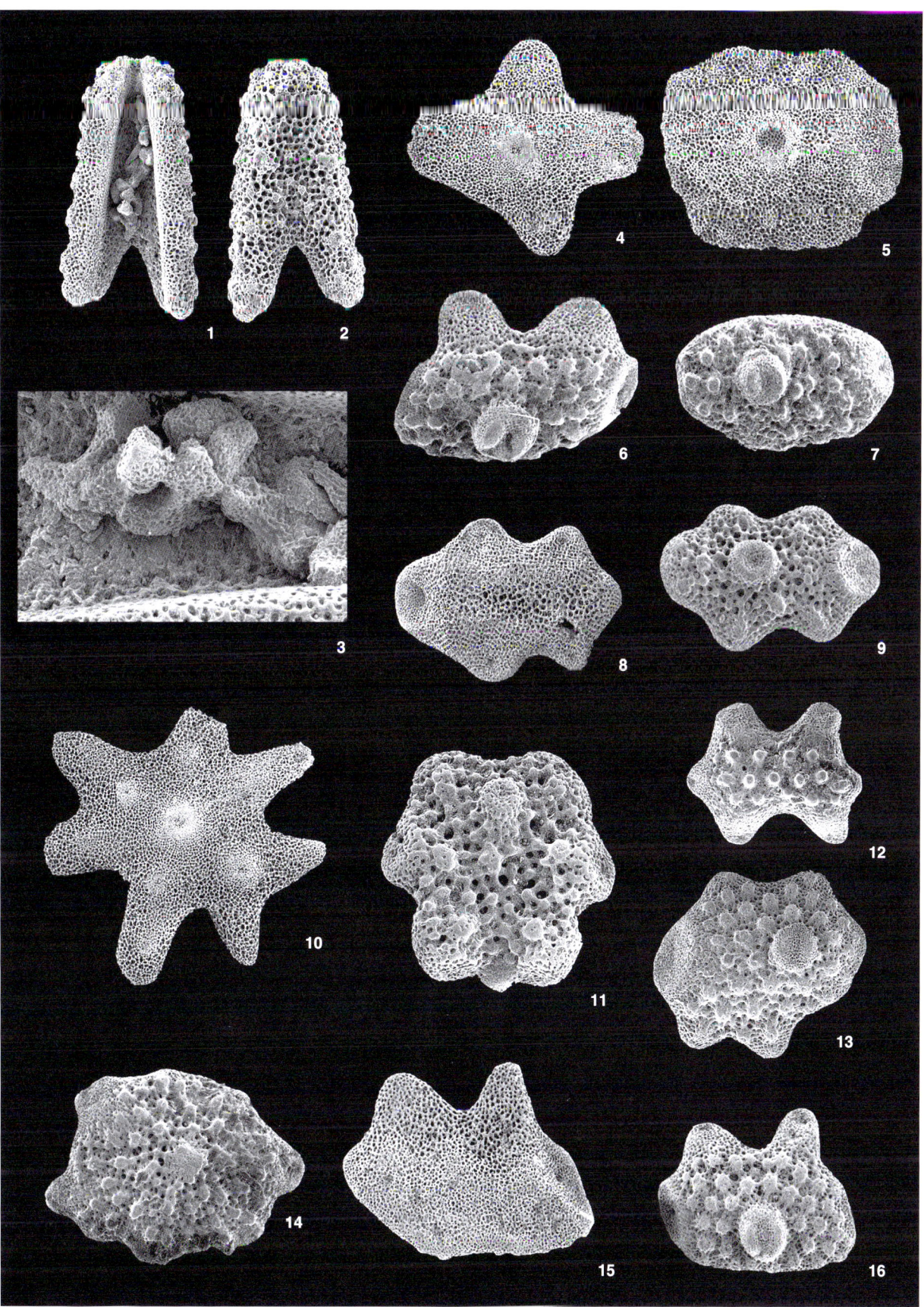

GALE, Fossil terminasterid and Recent zoroasterids

Acknowledgements. I wish to thank Andrew B. Smith (NHM) for his encouragement to undertake this study and for many helpful discussions, also to Peter Forey (NHM) for advice on cladistic topics. Ian Slipper (University of Greenwich) spent many months taking high-quality SEM photographs that form the basis of the study, and I am exceedingly grateful to him for such perseverance and dedication. Further help with SEM work was provided by Simon Cragg and Christine Hughes (both University of Portsmouth). Paul Taylor (NHM) took additional SEM photographs for which I am very thankful. David Ward (Orpington) assisted hugely with making up figures and plates and provided essential IT help. Kitty Thomas (Cambridge) drew several figures and provided advice on presentation and cladistic applications. I am grateful to Dan Blake (University of Illinois) and Fred Hotchkiss (Vineyard Haven) for discussions about asteroid phylogeny, although I suspect they will disagree profoundly with my conclusions. Two referees provided thorough and insightful reviews that I have taken into account in revision of the MS. I am most grateful for the loan and donation of specimens from various museum collections especially David Billett (IOS), Gordon Patterson and Andrew Kubrovinic (NHM), the late Cynthia Ahearn and David Pawson (Smithsonian Institution, Washington) and Fiona Ware (Royal Scottish Museum). The work has been greatly helped by the generosity of Hans Hess (Binningen), who gave me his asteroid library. Thanks are due to Sergei Rozhnov (PIN), who arranged access to the unique specimen of *Calliasterella mira* in the PIN collections at Moscow. The macrophotography was undertaken by Phil Crabb and Phil Hurst (both NHM Photographic Unit) to whom I am most grateful.

Editor. John Jagt

REFERENCES

ADKINS, W. S. 1928. Handbook of Texas Cretaceous fossils. *University of Texas Bulletin*, **2838**, 1–385, 37 pls.

ALCOCK, A. 1893. Natural history notes from H.M. Indian Marine Survey Steamer 'Investigator', commander R. F. Hoskyn, R.N., commanding, 2. An account of the deep-sea collection made during the season of 1892–93. *Journal of the Asiatic Society of Bengal*, **62**, 169–184, pls 8–9.

BAKER, A. N., ROWE, F. W. E. and CLARK, H. E. S. 1986. A new class of Echinodermata from New Zealand. *Nature*, **321**, 862–864.

BELYAEV, G. M. 1974. A new family of abyssal starfishes. *Zoologicheskii Zhurnal*, **53**, 1502–1508. [In Russian].

—— 1990. Is it valid to isolate the genus *Xyloplax* as an independent class of echinoderms? *Zoologicheskii Zhurnal*, **69**, 83–96. [In Russian].

BIZZARINI, F., LAGHI, G. F., NICOSIA, U. and RUSSO, F. 1989. Distribuzione stratigrafia dei microcrinoidi (Echinodermata) nella Formazione di S. Cassiano (Triassico Superiore, Dolomiti): studio preliminare. *Atti de la Società Matematica e Naturalisti di Modena*, **120**, 1–14.

BLAINVILLE, H. M. D. DE. 1830. *Dictionnaire des sciences naturelles, suivi d'une biographie des plus celèbres naturalistes par plusieurs professeurs du Jardin du Roi, et des principales écoles, Zoophytes.* F. G. Levrault, Paris, 546 pp.

BLAKE, D. B. 1972. Sea star *Platasterias*: ossicle morphology and taxonomic position. *Science*, **176**, 306–307.

—— 1973. Ossicle morphology of some Recent asteroids and description of some West American fossil asteroids. *University of California Publications in Geological Sciences*, **104**, 1–59, pls 1–18.

—— 1981. A reassessment of the sea-star orders Valvatida and Spinulosida. *Journal of Natural History*, **15**, 375–394.

—— 1982. Somasteroidea, Asteroidea, and the affinities *of Luidia* (*Platasterias*) *latiradiata. Palaeontology*, **25**, 167–191.

—— 1983. Some biological controls on the distribution of shallow water sea stars (Asteroidea; Echinodermata). *Bulletin of Marine Science*, **33**, 703–712.

—— 1984. The Benthopectinidae (Asteroidea, Echinodermata) of the Jurassic of Switzerland. *Eclogae geologicae Helvetiae*, **77**, 631–647.

—— 1986. Some new post-Palaeozoic sea stars (Asteroidea: Echinodermata) and comments on taxon endurance. *Journal of Paleontology*, **60**, 1103–1119.

—— 1987. A classification and phylogeny of post-Palaeozoic sea stars (Asteroidea: Echinodermata). *Journal of Natural History*, **21**, 481–528.

—— 1988*a*. A first fossil member of the Ctenodiscidae (Asteroidea, Echinodermata). *Journal of Paleontology*, **62**, 626–631.

—— 1988*b*. Paxillosidans are not primitive asteroids: a hypothesis based on functional considerations. 309–314. *In* BURKE, R. D., MLADENOV, P. V., LAMBERT, P. and PARSLEY, R. L. (eds). *Echinoderm Biology, Proceedings of the Sixth International Echinoderm Conference, Victoria, Australia.* A. A. Balkema, Rotterdam, Brookfield, 818 pp.

—— 1989. Asteroidea: functional morphology, classification and phylogeny. 179–223. *In* JANGOUX, M. and LAWRENCE, J. M. (eds). *Echinoderm Studies*, 3. Balkema, Rotterdam, Brookfield, 383 pp.

—— 1990. Hettangian Asteriidae (Echinodermata: Asteroidea) from southern Germany: taxonomy, phylogeny and life habits. *Paläontologische Zeitschrift*, **64**, 103–123.

—— and ELLIOT, D. R. 2003. Ossicular homologies, systematics, and phylogenetic implications of certain North American Carboniferous asteroids (Echinodermata). *Journal of Paleontology*, **77**, 476–489.

—— and GUENSBURG, T. E. 1988. The water vascular system and functional morphology of Paleozoic asteroids. *Lethaia*, **21**, 189–206.

—— and HAGDORN, H. 2003. The Asteroidea (Echinodermata) of the Muschelkalk (Middle Triassic of Germany). *Paläontologische Zeitschrift*, **77**, 23–58.

—— and HOTCHKISS, F. H. C. 2004. Recognition of the asteroid (Echinodermata) crown group: implications of the ventral skeleton. *Journal of Paleontology*, **78**, 359–370.

—— and REID, R. III. 1998. Some Albian (Cretaceous) asteroids (Echinodermata) from Texas and their paleobiological implications. *Journal of Paleontology*, **72**, 512–532.

—— and ZINSMEISTER, W. J. 1979. Two new Cenozoic sea stars (Class Asteroidea) from Seymour Island, Antarctic Peninsula. *Journal of Paleontology*, **53**, 1145–1154.

——BIELERT, F. and BIELERT, U. 2006. New early crown-group asteroids (Echinodermata; Triassic of Germany) *Paläontologische Zeitschrift,* **80**, 284–295.

—— TINTORI, A. and HAGDORN, H. 2000. A new early crown group asteroid (Echinodermata) from the Norian (Triassic) of Northern Italy. *Rivista Italiana di Paleontologia e Stratigrafia,* **106**, 141–156.

BOCZAROWSKI, A. 2001. Isolated sclerites of Devonian non-pelmatozoan echinoderms. *Palaeontologia Polonica,* **59**, 3–220.

BRANDT, J. F. 1835. *Prodromus descriptionis Animalium ab H. Mertensio in orbis Terrareum circumnavigationoe observatorum,* Vol. 1. Petripoli, 68–72.

BUCKMAN, J. 1844. Descriptions of the new species of the fossils. *In* MURCHISON, R. I., (ed.). *Outline of the geology of the neighbourhood of Cheltenham.* 2nd edition. John Murray, London, 109 pp.

CHIA, F.-S. and AMERONGEN, H. 1975. On the prey-catching pedicellariae of a starfish, *Stylasterias forreri* (de Loriol). *Canadian Journal of Zoology,* **53**, 748–755.

——OGURO, C. and KOMATSU, M. 1993. Sea-star (asteroid) development. *Oceanography and Marine Biology, Annual Review,* **31**, 223–257.

CLARK, A. M. 1984. Notes on Atlantic and other Asteroidea. 4. Families Poraniidae and Asteropseidae. *Bulletin of the British Museum (Natural History), Zoology,* **19**, 3, 81.

—— 1989. An index of names of Recent Asteroidea – part 1: Paxillosida and Notomyotida. 225–347. *In* JANGOUX, M. and LAWRENCE, J. M. (eds). *Echinoderm studies,* Vol. 3. A. A. Balkema, Rotterdam, Brookfield, 347 pp.

—— 1993. An index of names of Recent Asteroidea – part 2: Valvatida. 187–366. *In* JANGOUX, M. and LAWRENCE, J. M. (eds). *Echinoderm studies,* Vol. 4. A. A. Balkema, Rotterdam, Brookfield, 367 pp.

—— 1996. An index of names of Recent Asteroidea – part 3: Velatida and Spinulosida. 183–250. *In* JANGOUX, M. and LAWRENCE, J. M. (eds). *Echinoderm studies,* Vol. 5. A. A. Balkema, Rotterdam, Brookfield, 250 pp.

—— and DOWNEY, M. E. 1992. *Starfishes of the Atlantic.* Chapman & Hall, London, New York, Tokyo, Melbourne, Madras, 794 pp., pls 1–118.

—— and MAH, C. H. 2001. An index of names of Recent Asteroidea – part 4: Forcipulatida and Brisingida. 229–347. *In* JANGOUX, M. and LAWRENCE, J. M. (eds). *Echinoderm studies,* Vol. 6. A. A. Balkema, Rotterdam, Brookfield, 347 pp.

—— and ROWE, F. W. E. 1971. *Monograph of the shallow water Indo-West Pacific echinoderms.* Trustees of the British Museum (Natural History), London, 238 pp., pls 1–31.

CLARK, H. E. S. 1963. The fauna of the Ross Sea. Part 3. Asteroidea. *New Zealand Department of Science and Industry Research Bulletin,* **151**, 9–84, pls 1–14.

CLARK, H. L. 1908. Some Japanese and East Indian echinoderms. *Museum of Comparative Zoology Bulletin,* **51**, 289–301.

—— 1921. The echinoderm fauna of Torres Strait : its composition and its origin. *Paper of the Department of Marine Biology, Carnegie Institute, Washington,* **10**, 1–223, pls 1–38.

DANIELSSEN, D. C. and KOREN, J. 1884. Asteroidea. *Den Norske Nordhavns Expedition, 1876 1878,* **11**, 1–119, Grøndahl and Sønner, Christiania, pls 1–15.

DÖDERLEIN, L. 1916. Über die Gattung *Oreaster* und Verwandte Abteilung für systematik, Geographie und Biologie der Tiere, **40**, 409–440.

—— 1920. Die Asteriden der Siboga-Expedition, II: Die Gattung *Luidia* und ihre Stammesgeschichte. *Siboga-Expeditie Monograph,* **46b**, 193–294.

DOWNEY, M. E. 1970. Zorocallida, new order, and *Doraster constellatus,* new genus ands species, with notes on the Zoroasteridae (Echinodermata, Asteroidea). *Smithsonian Contributions to Zoology,* **64**, 1–18.

—— 1986. Revision of the Atlantic Brisingida (Echinodermata: Asteroidea), with description of a new genus and family. *Smithsonian Contributions to Zoology,* **435**, 1–57.

DÜBEN, M. W. and KOREN, J. 1846. Öfversigt af Skandinaviens Echinodermer. *Kungliga Svenska Vetenskaps Akademiens Handlingar,* **1844**, 227–328, pls 6–11.

DURHAM, J. W. and ROBERTS, W. A. 1948. Cretaceous asteroids from California. *Journal of Paleontology,* **22**, 432–439.

ERWIN, D. H. 1993. *The great Paleozoic crisis: life and death in the Permian.* Columbia University Press, New York, 327 pp.

ETHERIDGE, R. 1898. On the occurrence of a starfish in the Upper Silurian series of Bowning, New South Wales. *Records of the Australian Museum,* **3**, 128–129.

EYLERS, J. P. 1976. Aspects of skeletal mechanics of the starfish *Asterias forbesii. Journal of Morphology,* **149**, 353–368.

FELL, H. B. 1954. New Zealand fossil Asterozoa, 3 *Odontaster priscus* sp. nov. from the Jurassic. *Transactions of the Royal Society of New Zealand,* **82**, 817–819.

—— 1963. The phylogeny of sea-stars. *Philosophical Transactions of the Royal Society London, Series B,* **246**, 381–435.

—— and CLARK, H. E. 1959. *Anareaster,* a new genus of Asteroidea from Antarctica. *Transactions of the Royal Society of New Zealand,* **87**, 185–187.

FISHER, W. K. 1905. New starfishes from deep water off California and Alaska. *Bureau of the United States Fisheries,* **24**, 291–320.

—— 1906. The starfishes of the Hawaiian Islands. *Bulletin of the United States Fisheries Commission,* **23**, 987–1130, pls 1–49.

—— 1910. New starfishes from the North Pacific. 1. Phanerozonia. 2. Spinulosida. *Zoologischer Anzeiger,* **35**, 545–553, 568–574.

—— 1911. Asteroidea of the North Pacific and adjacent waters, 1. Phanerozonia and Spinulosa. *Bulletin of the United States National Museum,* **76**, xiii + 1–xiii 420, pls 1–122.

—— 1913. Four new genera and fifty-eight new species of starfishes from the Phillipine Islands, Celebes, and the Moluccas. *Proceedings of the United States National Museum,* **43**, 599–648.

—— 1916. New East Indian starfishes. *Proceedings of the Biological Society of Washington,* **29**, 27–36.

—— 1919a. Starfishes of the Phillippine seas and adjacent waters. *Bulletin of the United States National Museum,* **3**(100), 1–547, pls 1–156.

—— 1919b. North Pacific Zoroasteridae. *Annals and Magazine of Natural History,* **3** (9), 347–353.

—— 1923. A preliminary synopsis of the Asteriidae, a family of sea-stars. *Annals and Magazine of Natural History*, **12** (9), 247–258, 595–607.

—— 1928. Asteroidea of the North Pacific and adjacent waters, Part 2. Forcipulata (Part). *Bulletin of the United States National Museum*, **76**, 245 pp, 81 pls.

—— 1930. Asteroidea of the North Pacific and adjacent waters, Part 3. Forcipulata (Concluded). *Bulletin of the United States National Museum*, **76**, 356 pp, 93 pls.

—— 1940. Asteroidea. *Discovery Report*, **20**, 69–306, pls 1–23.

FORBES, E. 1839. On the Asteriadae of the Irish Sea. *Memoir of the Wernerian Society of Edinburgh*, **8**, 114–129, 2 pls.

—— 1841. *A history of British starfish and other animals of the class Echinodermata.* John Van Voorst, London, 267 pp.

—— 1848. On the Asteridae found fossil in British Strata. *Memoirs of the Geological Survey of Great Britain, British Organic Remains*, Decade **2**, 457–482.

—— 1850. *In* DIXON, F. (ed). *The geology and fossils of the Tertiary and Cretaceous formations of Sussex.* Longman, Brown, Green and Longman, London, 422 pp, 50 pls.

—— 1856. *Solaster moretonensis. Memoirs of the Geological Survey of Great Britain, British Organic Remains*, Decade **5**, 1–3, pl. 1.

GALE, A. S. 1987. Phylogeny and classification of the Asteroidea (Echinodermata). *Zoological Journal of the Linnean Society*, **89**, 107–132.

—— 2005. *Chrispaulia*, a new genus of mud star (Asteroidea, Goniopectinidae) from the Cretaceous of England. *Geological Journal*, **40**, 383–397.

GRAY, J. 1840. A synopsis of the genera and species of the Class Hypostoma (*Asterias* Linnaeus). *Annals and Magazine of Natural History*, **6**, 175–184, 275–290.

—— 1847. Descriptions of some new genera and species of Asteriadae. *Annals and Magazine of Natural History*, **20**, 193–204.

GREGORY, J. W. 1899. On *Lindstromaster* and the classification of the palaeasterids. *Geological Magazine*, **6**, 341–354.

HAYASHI, R. 1957. A note of a Japanese sea-star, *Astropecten ludwigi* de Loriol. *Journal of the Faculty of Science of Hokkaido Imperial University, Zoology*, **13**, 37–45.

—— 1974. A new sea-star from Japan, *Asterina minor* sp. nov. *Proceedings of the Japanese Society for Systematic Zoology*, **10**, 41–44.

HEDDLE, D. 1967. Versatility of movement and the origin of the asteroids. *Symposium of the Zoological Society of London*, **20**, 125–141.

—— 1995. The descent of the Asteroidea and the reaffirmation of paxillosidan primitiveness. 179–183. *In* EMSON, R., SMITH, A. B. and CAMPBELL, A. (eds). *Echinoderm Research 1995.* Proceedings of the Fourth European Echinoderms Colloquium, London, UK. A.A. Balkema, Amsterdam, Brookfield, 314 pp.

HESS, H. 1955. Die fossilen Astropectiniden (Asteroidea). *Schweizerische Paläontologische Abhandlungen*, **71**, 1–113, pls 1–4.

—— 1968. Ein neuer Seestern (*Pentasteria longispina* n. sp.) aus den Effingerschichten des Weissensteins (Kt. Solothurn). *Eclogae geologicae Helvetiae*, **61**, 607–614.

—— 1970. Schlangensterne und Seesterne aus dem oberen Hauterivien "Pierre jaune" von St-Blaise bei Neuchâtel. *Eclogae geologicae Helvetiae*, **63**, 1069–1091.

—— 1971. Neue Funde des Seesterns *Terminaster cancriformis* (Quenstedt) aus Callovien und Oxford von England, Frankreich und der Schweiz. *Eclogae geologicae Helvetiae*, **67**, 647–659.

—— 1972. Eine Echinodermen-Fauna aus dem mittleren Dogger des Aargauer Juras. *Schweizerische Paläontologische Abhandlungen*, **92**, 1–87, pls 1–23.

—— 1974. Neue Funde des Seesterns *Terminaster cancriformis* (Quenstedt) aus Callovien und Oxford von England, Frankreich und der Schweiz. *Eclogae geologicae Helvetiae*, **67**, 647–659.

—— 1987. Neue Seesternfunde aus dem Dogger des Schweizer Juras. *Eclogae geologicae Helvetiae*, **80**, 907–918.

HESSELBO, S. P. 2000. Late Triassic and Jurassic: disintegrating Pangaea. 314–338. *In* WOODCOCK, N. and STRACHAN, R. (eds). *The geological history of the British Isles.* Blackwell's, Oxford, 423 pp.

HOTCHKISS, F. H. C. and CLARK, A. M. 1976. Restriction of the family Poraniidae, sensu Spencer & Wright, 1966 (Echinodermata: Asteroidea). *Bulletin of the British Museum of Natural History (Zoology)*, **30**, 263–268, pls 1–3.

HUDSON, G. H. 1912. A fossil starfish with ambulacral covering plates. *Ottowa Naturalist*, **26**, 21–26, 45–52, pls 1–3.

IVES, J. E. 1888. On two new species of starfish. *Proceedings of the Academy of Natural Sciences of Philadelphia*, **1888**, 421–424.

JAGT, J. W. M. 1991. Early Miocene luidiid asteroids (Echinodermata, Asteroidea) from Winterswijk-Miste (Netherlands). *Contributions to Tertiary and Quaternary Geology*, **28**, 35–43.

JANGOUX, M. 1982. Digestive systems: Asteroidea. 235–279. *In* JANGOUX, M. and LAWRENCE, J. M. (eds). *Echinoderm nutrition.* A. A. Balkema, Rotterdam, Brookfield, 654 pp.

—— and LAMBERT, A. 1987. Étude comparative des pédicellaires des astérides (échinodermes). *Bulletin de la Société scientifique et naturaliste de l'Ouest de France, supplément hors-série*, 47–56.

JANIES, D. 2001. Phylogenetic relationships of extant echinoderm classes. *Canadian Journal of Zoology*, **79**, 1232–1250.

—— and McEDWARD, L. 1992. Highly derived coelomic and water-vascular morphogenesis in a starfish with pelagic direct development. *Biological Bulletin*, **185**, 56–76.

—— and MOOI, R. 1998. *Xyloplax* is an asteroid. 311–316. *In* CARNEVALI, M. D. C. C. and BONAROSO, F. *Echinoderm Research.* Proceedings of the Fifth European Conference on Echinoderms, Milan, Italy. A. A. Balkema, Rotterdam, Brookfield, 449 pp.

JONES, T. R. 1887. *Notes on some Silurian Ostracods from Gothland.* Kongliga Boktryckeriet, P. A. Nordstedt and Sønner, Stockholm, 8 pp.

JONES, D. S. and PORTELL, R. W. 1988. Occurrence and biogeographic significance of *Heliaster* (Echinodermata, Asteroidea) from the Pliocene of southwest Florida. *Journal of Paleontology*, **62**, 126–132.

KACZMARSKA, G. 1987. Asteroids from the Korytnica Basin (Middle Miocene; Holy Cross Mountains, Central Poland). *Acta Geologica Polonica*, **37**, 131–144.

KANO, Y. T., KOMATSU, M. and OGURO, C. 1974. Notes on the development of the sea-star *Leptasterias ochotensis similispinis*, with special reference to skeletal system. *Proceedings of the Japanese Society for Systematic Zoology*, **10**, 45–53.

KESLING, R. V. 1967. *Neoplaeaster enigmaticus*, new starfish from upper Mississippian Paint Creek Formation in Illlinois. *Contributions from the Museum of Paleontology, University of Michigan*, **21**, 73–85.

—— 1969. Three Permian starfish from Western Australia and their bearing on the revision of the Asteroidea. *Contributions from the Museum of Paleontology, University of Michigan*, **22**, 361–376.

—— 1982. *Arkonaster*, a new multi-armed starfish from the Middle Devonian Arkona Shale of Ontario. *Contributions from the Museum of Paleontology, University of Michigan*, **26**, 83–115.

—— and STRIMPLE, H. L. 1966. *Calliasterella americana*, a new starfish from the Pennsylvanian of Illinois. *Journal of Paleontology*, **40**, 1157–1166.

KITCHING, I. J., FOREY, P. L., HUMPHRIES, C. J. and WILLIAMS, D. M. 1999. *Cladistics*. Oxford University Press, Oxford, New York, 228 pp.

KNOTT, K. E. and WRAY, G. A. 2000. Controversy and consensus in asteroid systematics: new insights to ordinal and familial relationships. *American Zoologist*, **40**, 382–393.

KOEHLER, R. 1906. Echinodermes (Stéllérides, Ophiures, et Echinides). *Expéditiion antarctique Francaise (1903–1905) commandée par Dr. Jean Charcot*. Sciences naturelles: Documents scientifique, 1–41, pls 1–4.

—— 1912. Echinodermes (Stéllérides, Ophiures, et Echinides). *Deuxieme Expédition Antarctique Francaise (1908–1910)*. Science Naturelle, 1–272, pls 1–16.

—— 1920. Echinodermata Asteroidea. *Scientific Reports of the Australian Antarctic Expedition*, **C8**, 1–308, pls 1–75.

KOMATSU, M. 1975. On the development of the sea-star *Astropecten latespinosus* Meissner. *Biological Bulletin*, **148**, 49–59.

—— KANO, Y. T., YOSHIZAWA, H., AKABANE, S. and OGURO, C. 1979. Reproduction and development of the hermaphroditic sea-star *Asterina minor* Hayashi. *Biological Bulletin*, **157**, 258–274.

LAFAY, B., SMITH, A. B. and CHRISTEN, R. 1995. A combined morphological and molecular approach to the phylogeny of asteroids (Asteroidea: Echinodermata). *Systematic Biology*, **44**, 190–208.

LAMARCK, J. B. P. A. 1816. Stellerides. *Histoire Naturelle des animaux sans vertebres*, First edition, Vol. 2. Déterville & Verdiére, Paris, 522–568.

LAMBERT, A., DE VOS, L. and JANGOUX, M. 1984. Functional morphology of the pedicellariae of the asteroid *Marthasterias glacialis* (Echinodermata). *Zoomorphology*, **104**, 122–130.

LIEBERKIND, I. 1926. *Ctenodiscus australis* Lütken, a brood-protecting asteroid. *Videnskabernes Meddelelser fra det Dansk Naturhistorisk Forening*, **82**, 183–196.

LINNAEUS, C. 1758. *Systema Naturae*, Tenth edition. Holmiae, Verdiére, Paris,, 824 pp.

LORIOL, P. DE. 1873. Description de quelques Astérides du terrain néocomien des environs de Neuchatel. *Mémoires de la Société Scientifique naturelle de Neuchatel*, **4**, part 2.

—— 1887. Notes pour servir a l'étude des Echinodermes 2. *Receuil Zoologique Suisse*, **4**, 401.

—— 1897a. Notes sur quelques étoiles de mer de la Jurassique du sud est de France. *Revue Suisse de Zoologie*, **5**, 177, pl. 8.

—— 1897b. Notes pour servir a l'étude des Echinodermes. 5. *Memoires de la Societé Physique et d'Histoire Naturelle, Geneva*, **32** (9), 21, pl. 3.

—— 1899. Uber einen neuen fossilen Seesterne. *Mitteilungen Grossherzoglich Badischen Geologischen Landesanstalt*, **4**, 5–10.

LUDWIG, H. 1907. Diagnosen neuer Tiefsee Seesterne aus der Familie der Porcellanasteridae. *Zoologischer Anzeiger*, **31**, 312–319.

—— 1910. Notomyota, eine neue Ordnung der Seesterne. *Sitzungsberichte der Koniglch Preussischen Akademie der Wissennschaften zu Berlin*, **23**, 435–466.

LÜTKEN, J. 1871. Fortsatte kritiske og beskrivende Bidrag til Kundskab om Sostjernerne (Asteriderne). *Videnskabelige Meddelelser fra Dansk naturhistorisk Forening i Kobenhavn*, **238** (14), 238 pp.

MacBRIDE, E. W. 1921. Echinoderm larvae and their bearing on evolution. *Nature*, **108**, 529–530.

McEDWARD, L. R. and JANIES, D. A. 1993. Life cycle evolution in asteroids: what is a larva? *Biological Bulletin*, **184**, 255–268.

McKINNEY, M. L. and McNAMARA, K. J. 1991. *Heterochrony: the evolution of ontogeny*. Plenum Press, New York, 437 pp.

McKNIGHT, D. G. 1975. Classification of asteroids and somasteroids (Asterozoa: Echinodermata). *Journal of the Royal Society of New Zealand*, **5**, 13–19.

MADSEN, F. J. 1961. The Porcellanasteridae: a monographic revision of an abyssal group of sea-stars. *Galathea Report*, **4**, 33–174.

—— 1966. The Recent sea-star *Platasterias* and the fossil Somasteroidea. *Nature*, **209**, 1367.

MAH, C. L. 2000. Preliminary phylogeny of the forcipulatacean Asteroidea. *American Zoologist*, **40**, 375–381.

—— 2003. *Astrosarkus idipi*, a new Indo-Pacific genus and species of Oreasteridae (Valvatida, Asteroidea) displaying extreme endoskeletal reduction. *Bulletin of Marine Science*, **73**, 685–698.

—— 2006. A new species of *Xyloplax* (Echinodermata: Asteroidea: Concentricocycloidea) from the northeast Pacific: comparative morphology and a reassessment of phylogeny. *Invertebrate Zoology*, **125**, 136–153.

—— 2007. Phylogeny of the Zoroasteridae (Zorocallina, Forcipulatida): evolutionary events in deep sea Asteroidea displaying Palaeozoic features. *Zoological Journal of the Linnean Society*, **150**, 177–210.

MARENZELLER, E. [von] 1893. Berichte der Commission für Erforschung des östlichen Mittelmeer. Zoologische Ergebnisse, 1. Echinodermen, gesammelt 1890, 1891 & 1892. *Denkschrifte der Akademie der Wissenschaften zu Wien*, **60**, 1–24.

MASATOSHI, Y. 1987. Asteroids from the Miocene of Tokai. *Publication of the Tokai Fossil Society*, **31**, 1–23. [In Japanese].

MATSUBARA, M., KOMATSU, M. and WADA, H. 2004. Close relationship between *Asterina* and Solasteridae (Asteroidea) supported by both nuclear and mitochondrial gene molecular phylogenies. *Zoological Science*, **21**, 785–793.

—— ARAKI, T., ASAKAWA, S., YOKOBORI, S., WATANABE, K. and WADA, H. 2005. The phylogenetic status of Paxillosida (Asteroidea) based on complete mitochondrial DNA sequences. *Molecular Phylogenetics and Evolution*, **36**, 598–605.

MEISSNER, M. 1892. Asteriden gesammelt von Herrn Stabsartz Dr. Sander auf S.M.S Prinz Adalbert. *Archive fuer Naturgeschichte*, **62** (1), 91–108.

MOOI, R. and DAVID, B. 2000. What a new model of skeletal homologies tell us about asteroid evolution. *American Zoologist*, **40**, 326–339.

MORTENSEN, T. 1921. *Studies of the development and larval forms of echinoderms*. C. A. Reitzel, Copenhagen, 261 pp., 33 pls.

—— 1925. Echinoderms of New Zealand and the Auckland-Campbell Islands. 3-5 Asteroidea, Holothuroidea, Crinoidea. Zoogeographical remarks on the echinoderm fauna of New Zealand and the Auckland-Campbell Islands. *Videnskabelige Meddelelser fra Dansk naturhistorisk Forening i Kobenhavn*, **79**, 261–420, pls 12–14.

MÜLLER, J. and TROSCHEL, F. H. 1840. Über die Gattungen der Asteriden. *Wiegmanns Archiv für Naturgeschichte*, **6**, 318–326, 328–368.

—— —— 1842. *System der Asteriden*. F. Vieweg und Sohn, Braunschweig, xx + 134 pp., 12 pls.

MÜLLER, O. F. 1776. *Zoologiae Danicae Prodromus, seu Animalium Daniae et Norvegiae indegenarium, etc.*. Havniae, Hallager, 274 pp.

MÜNSTER, G. von. 1843. *Asterias weissmanni, Beitrage zur Petrefactenkunde*, **6**, 78.

MURDOCH, J. 1885. Marine invertebrates. *In Report of the international Polar Expedition to Point Barrow, Alaska*. Government Printing Office, Washington DC, 693 pp.

NANCE, J. M. and BRAITHWAITE, L. F. 1972. The function of mucus secretions in the cushion star *Pteraster tesselatus* Ives. *Journal of Experimental Marine Biology and Ecology*, **40**, 259–266.

NICHOLS, D. 1969. *Echinoderms*, Fourth revised edition. Hutchinson University Library, London, 192 pp.

OGURO, C. 1989. Evolution of development and larval types in asteroids. *Zoological Science*, **6**, 199–210.

—— KOMATSU, M. and KANO, K. T. 1976. Development and metamorphosis of the sea-star *Astropecten scoparius* Valenciennes. *Biological Bulletin*, **151**, 560–573.

O'LOUGHLIN, P. M. and WATERS, J. M. 2004. A molecular and morphological revision of genera of Asterinidae (Echinodermata: Asteroidea). *Memoirs of Museum Victoria*, **61**, 1–40.

PATTERSON, C. and SMITH, A. B. 1987. Is the periodicity of extinctions a taxonomic artefact? *Nature*, **330**, 248–251.

PENNANT, T. 1777. *British Zoology*, Fourth edition, Vol. 4. B. White, London, 154 pp.

PEREIRA, P., CACHÃO, M. and MARQUES DA SILVA, C. 2003. Asteroidea (Echinodermata) do Miocénico da Bacia do Baixo Tejo-Sado. *Ciências da Terra (UNL), numero especial*, **5**, A106–A109.

PERRIER, E. 1875. *Revision de la collection de Stéllerides du Muséum d'Historie Naturelle de Paris*. Reinwald, Paris, 384 pp.

—— 1881. Description sommaire des especes nouvelles d'Astéries. *Bulletin of the Museum of Comparative Zoology, Harvard*, **9**, 1–31.

—— 1884. Mémoire sur les étoiles de mer recueillis dans la Mer des Antilles et le Golfe de Mexique. *Nouvelles Archives du Muséum d'Histoire Naturelle de Paris*, **6** (2), 127–276.

—— 1885. Sur les stellerides recueillis durant la mission du Talisman. *Comptes Rendus Hebdomadaires des Séances de l'Academie des Sciences*, Paris, **101**, 884–887.

—— 1893. *Traité de Zoologie*, **1** (2), Paris, 864 pp.

—— 1894. Stéllerides. *Expédition Scientifique du Travailleur-Talisman*, **3**, 1–143, pls 1–26.

QUENSTEDT, F. A. 1874–1876. *Petrifactenkunde Deutschlands*. 4. Band (Echinodermen) (Asteriden und Encriniden), Fues, Leipzig, 742 pp.

PHILIPPI, R. A. 1837. Uber die mit *Asterias auranciaca* verwanten und verwechselten Asterien der Sicilianschen Kuste. *Archiv fuer Naturgeschichte*, **3**, 193–194.

RASMUSSEN, H. WIENBERG. 1972. Lower Tertiary Crinoidea, Asteroidea and Ophiuroidea from northern Europe and Greenland. *Det Kongelige Danske Videnskabernes Selskab, Biologiske Skrifter*, **19**, 1–83, pls 1–14.

RETZIUS, A. J. 1783. Anmaerkningar vid Asteriae Genus. *K. Konglige Svenska Vetenskaps-Akademiens Handlingar*, **4**, 234–244.

—— 1805. *Dissertatio Sistens Species Cognitas Asteriarum*. Lund, 37 pp.

RISSO, A. 1826. Echinodermes. 267–291. *In Histoire naturelle des principales productions de l'Europe meridionale et particulierement de celles des environs de Nice et des Alpes Maritimes*, Vol. 5. Levrault, Paris, 439 pp.

ROBERTS, M. P. and CAMPBELL, A. C. 1988. Functional anatomy of pedicellariae from *Asterias rubens* L. 725–733. *In* BURKE, R. D., MLADENOV, P. V., LAMBERT, P. and PARSLEY, R. L. (eds). *Echinoderm Biology, Proceedings of the Sixth International Echinoderm Conference, Victoria, Australia*. Balkema, Rotterdam, Brookfield, 818 pp.

ROUX, M. 1970. Introduction à l'étude des microstructures des tiges de crinoïdes. *Geobios*, **3**, 79–98.

—— 1971. *Recherches sur la microstructure des pédoncules de crinoïdes post-Paléozoiques*. Travaux du Laboratoire Paléontologique de la Faculté de Science, Université de Paris, Orsay, 86 pp.

ROWE, F. W. E., BAKER, A. N. and CLARK, H. E. S. 1988. The morphology, development and taxonomic status of *Xyloplax* Baker, Rowe and Clark (1986) (Echinodermata: Concentricycloidea), with a description of a new species. *Proceedings of the Royal Society London, Series B*, **233**, 431–459.

SALTER, J. W. 1857. On some new Palaeozoic star-fishes. *Annals and Magazine of Natural History, 2nd series*, **20**, 321–334.

SARS, G. O. 1875. *On some remarkable forms of animal life from the great deeps [sic] off the Norwegian coast, II. Researches on the structure and affinity of the genus* Brisinga. A. W. Brøgger, Christiania, 112 pp.

SARS, M. 1861. *Oversigt af Norges Echinodermer*. Christiana, Norway, 160 pp.

SCHÄFER, W. 1962. *Aktuo-Paläontologie nach Studien in der Nordsee*. Waldemar Kramer, Frankfurt am Main, 666 pp.

SCHÖNDORF, F. 1909. Die Asteriden des russischen Karbon. *Palaeontographica*, **45**, 323–338, pls 23–24.

SCHUCHERT, C. 1914. Revision of the Paleozoic Stelleroidea with special reference to North American Asteroidea. *Bulletin of the United States National Museum*, **88**, 311 pp.

SHACKLETON, J. D. 2005. Skeletal homologies, phylogeny and classification of the earliest asterozoan echinoderms. *Journal of Systematic Palaeontology*, **3**, 29–114.

SHICK, J. M., EDWARDS, K. C. and DEARBORN, J. H. 1981. Physiological ecology of the deposit-feeding sea-star *Ctenodiscus crispatus*: ciliated surfaces and animal-sediment interactions. *Marine Ecology Progress Series*, **5**, 165–184.

SLADEN, W. P. 1882. Description of *Mimaster*, a new genus of Asteroidea from the Faroe Channel. *Transactions of the Royal Society of Edinburgh*, **30**, 579–584.

—— 1883. The Asteroidea of H.M.S. Challenger Expedition. (Preliminary notices). 2. Astropectinidae. *Journal of the Linnean Society of London, Zoology*, **17**, 214–269.

—— 1885. Asteroidea. *In* WYVILLE THOMSON, C. and MURRAY, J. (eds). *Reports of the Scientific Results of the Voyage of the Challenger, 1873–1876. Zoology*, **1** (2), 607–617.

—— 1889. The Asteroidea. *Report of the Scientific Results of the Voyage of the H.M.S. Challenger, Zoology*, **30**, 935. pp., pls 1–118.

SMITH, A. B. 1980. Stereom microstructure of the echinoid test. *Special Papers in Palaeontology*, **25**, 1–81.

—— 1988. To group or not to group; the taxonomic position of *Xyloplax*. 17–23. *In* BURKE, R. D., MLADENOV, P. V., LAMBERT, P. and PARSLEY, R. L. (eds). *Echinoderm Biology, Proceedings of the Sixth International Echinoderm Conference, Victoria, Australia*. A.A. Balkema, Rotterdam, Brookfield, 818 pp.

—— 1990. Echinoid evolution from the Triassic to the Lower Jurassic. *Cahiers Université Catholique de Lyon, Séries Science*, **3**, 79–117.

—— 1994. *Systematics and the fossil record: documenting evolutionary patterns*. Blackwell Scientific Publications, Oxford, 223 pp.

—— 2001. Large scale heterogeneity of the fossil record: implications for biodiversity studies. *Philosophical Transactions of the Royal Society London, Series B*, **356**, 351–367.

—— and McGOWAN, A. J. 2007. The shape of the Phanerozoic marine palaeodiversity curve: how much can be predicted from the sedimentary rock record of western Europe. *Palaeontology*, **50**, 765–774.

—— and TRANTER, T. H. 1985. *Protremaster*, a new Lower Jurassic genus of asteroid from Antarctica. *Geological Magazine*, **122**, 351–359.

SMITH, E. A. 1876. Descriptions of species of Asteriidae and Ophiuridae from Kergulen Island. *Annals and Magazine of Natural History*, **17** (4), 105–113.

SPENCER, W. K. 1905. A monograph on the British fossil Echinodermata from the Cretaceous Formations. *Palaeontographical Society (Monograph)*, Volume 2. The Asteroidea and Ophiuroidea. Part 3, 67–90, pls 17–26.

—— 1913. The evolution of the Cretaceous Asteroidea. *Philosophical Transactions of the Royal Society London, Series B*, **204**, 99–177.

—— 1915. *Archaster patersoni*, n. sp. A new South African fossil starfish. *Records of the Albany Museum*, **3**, 65–69.

—— and WRIGHT, C. W. 1966. Asterozoans. U4–U107. *In* MOORE, R. C. (ed.). *Treatise on Invertebrate Paleontology, Echinodermata*, **3** (1). Geological Society of America, Boulder, Colorado and The University of Kansas Press, Lawrence, Kansas, 366 pp.

STIMPSON, W. 1857. On the Crustacea and Echinodermata of the Pacific shores of North America. *Boston Journal of Natural History*, **6**, 444–532.

STUDER, T. 1883. Uber die Asteriden welche wahrend der Reise S.M.S. *Gazelle gesamelt. Sitzungberichte der Gesellschaft Naturforchender Freunde zu Berlin.*, **1883** (8), 128–132.

SUKARNO and JANGOUX, M. 1977. Révision du genre *Archaster* Muller et Trotschel. *Revue Zoologique Afrique*, **91**, 817–844, pls 4–6.

SUMIDA, P. Y. G., TYLER, P. A. and BILLETT, D. S. M. 2001. Early juvenile development of deep-sea asteroids of the NE Atlantic Ocean, with notes on juvenile bathymetric distributions. *Acta Zoologica*, **82**, 11–40.

SUTTON, M. D., BRIGGS, D. E. G. and SIVETER, D. J. 2005. A starfish with three-dimensionally preserved soft parts from the Silurian of England. *Proceedings of the Royal Society London, Series B*, **272**, 1001–1006.

TRAUTSCHOLD, H. 1879. Die Kalkbrüche von Mjatschkowa, Teil 3. *Nouveaux Mémoires de la Société imperiale des Naturalistes de Moscou*, **14**, 101–108, pls 1–7.

TURNER, R. L. and DEARBORN, J. H. 1972. Skeletal morphology of the mud star, *Ctenodiscus crispatus* (Echinodermata: Asteroidea). *Journal of Morphology*, **138**, 239–262.

VALETTE, A. 1929. Note sur quelques stellerides Jurassiques. *Travaux du Laboratoire de Géologie de La Faculté des Sciences de Lyon*, **16**, 7–39, pls 1–5.

VERRILL, A. E. 1866. On the polyps and echinoderms of New England with descriptions of new species. *Proceedings of the Boston Society of Natural History*, **10**, 333–375.

—— 1870. Notes on Radiata in the Museum of Yale College. 1. Descriptions of new starfishes from New Zealand. 2. Notice on the echinoderms of Panama and the West coast of America. 3. On the geographical distribution of echinoderms of the Pacific coast of America. *Transactions of the Connecticut Academy of Arts and Science*, **1** (2), 247–251, 251–322, 323–351.

—— 1880. Notice of recent additions to the marine fauna of the eastern coast of North America. 8. *American Journal of Science*, **19**, 137–140.

—— 1884. Notice of the remarkable marine fauna occupying the outer banks off the southern coast of New England. 9, 10. *American Journal of Science*, **28**, 213–220, 378–384.

—— 1899. Revision of certain genera and species of starfishes with descriptions of new forms. *Transactions of the Connecticut Academy*, **10**, 145–234.

—— 1914. Monograph of the shallow-water starfishes of the North Pacific coast from the Arctic Ocean to California. *Harriman Alaska Series, United States National Museum*, **14**, 408 pp., 110 pls.

VICKERY, M. S. and McCLINTOCK, J. B. 2000. Comparative morphology of tube feet among the Asteroidea: phylogenetic implications. *American Zoologist*, **40**, 355–364.

VIGUIER, C. 1879. Anatomie comparée du squelette des stellérides. *Archives de Zoologie Expérimentale et Générale*, **7**, 33–250.

VILLIER, L., CHARBONNIER, S. and RIOU, B. 2009. Sea stars from Middle Jurassic Lagerstätte of La Voulte-sur-Rhône (Ardèche, France). *Journal of Paleontology*, **83**, 389–398.

—— KUTSCHER, M. and MAH, C. L. 2004*a*. Systematics and palaeoecology of middle Toarcian Asteroidea from the "Seuil du Poitou", western France. *Geobios*, **37**, 807–825.

—— BLAKE, D. B., JAGT, J. W. M. and KUTSCHER, M. 2004*b*. A preliminary phylogeny of the Pterasteridae (Echinodermata, Asteroidea) and the first fossil record: Late Cretaceous of Belgium and Germany. *Paläontologische Zeitschrift*, **78**, 281–299.

—— BRETON, G., MARGERIE, P. and NÉRAUDEAU, D. 2004*c*. *Manfredaster* gen. nov. *cariniferus* sp. nov., un astéride original du Coniacien de Seine-Maritime et révision systématique de la famille des Stauranderasteridae (Echinodermata, Asteroidea). *Bulletin trimestriel de la Société Géologique de Normandie et des Amis du Muséum du Havre*, **90**, 29–41.

WADA, H., KOMATSU, M. and SATOH, N. 1996. Mitochondrial rDNA phylogeny of the Asteroidea suggests the primitiveness of the Paxillosida. *Molecular Phylogenetics and Evolution*, **6**, 97–106.

WALBRAN, P. D., HENDERSON, R. A., JULL, A. J. T. and GEAD, M. J. 1989. Evidence from sediments of long-term *Acanthaster planci* predation on corals on the Great Barrier Reef. *Science*, **245**, 847–850.

WOOD-MASON, J. and ALCOCK, A. 1891. Natural history notes from H.M. Marine Survey Steamer Investigator. Echinoderms. *Annals and Magazine of Natural History*, **8** (6), 427–443, 1 pl.

WRIGHT, T. 1862–1880. A monograph on the British fossil Echinodermata of the Oolitic Formations, Vol. II. The Asteroidea and Ophiuroidea. *Monograph of the Palaeontographical Society London*, **1862**, 1–154; **1863–1880**, 155–203.

WYVILLE THOMSON, C. 1873. *The Depths of the Sea*. Macmillan and Company, London, xx + 527 pp.

XANTUS, J. 1860. Descriptions of three new species of starfishes from Cape St Lucas. *Proceedings of the Academy of Natural Sciences in Philadelphia*, **1860**, 568.

ZARDINI, R. 1973. *Fossili di Cortina. Alante deglı echınodermı cassiani (Trias medio-superiore) della regione dolomitica attorno a Cortina d'Ampezzo*. Foto Ghedina, Cortina d'Ampezzo, 29 pp., pls 1–22.

APPENDIX: CHARACTER LIST

General morphology

1. Arms 5 (0), 5–10 (1), >10 (2)
2. Arms very short, disc large (R:r 1–2) (0), arms moderately produced, disc medium sized (R:r 2–4) (1), arms elongated, disc small (R:r > 4) (2)
3. Interbrachial arcs angled (0), curved (1)
4. Arm cross-section subcylindrical (0), convex abactinally, flat actinally (1), rectangular (2)
5. Spines constructed of fine labyrinthic stereom (0), shaft composed partially or completely of glassy, elongated trabeculae (1)
6. Spines elongated (0), short, granular, forming polygonal tesselation (1)
7. Calcified interradial septa absent (0), form a pillar-like structure articulating with odontophore (1), form an interradial partition contiguous with actinal and abactinal ossicles (2)
8. Interradial chevron ossicles (*ico*) absent (0), present (1)
9. Interradial grooves (*ig*) absent (0), present (1)
10. Tube feet tips pointed (0), flat (1)
11. Tube feet without sucker (0) with sucker (1)
12. Anus and rectal caecae absent (0), present (1)
13. Stomach noneversible, no retractor (0), eversible with retractor (1)
14. Tiedemann's pouches absent (0), present (1)
15. Pyloric complex small (0), large, well developed (1)

Abactinals, marginals, actinals

16. Abactinals differentiated into larger primary ossicles and smaller secondaries (0), all of similar small size (1), reduced to small ossicles in abactinal membrane (2)

17. Ratio of the distance between the inner margin of the superomarginal and the distance between the centrale and madreporite (PM) >1 (0), <1 (1)
18. Madreporite separate (0), fused with primary interradial and adjacent paired interradials (1)
19. Abactinals tessellate (0), reticulate (1)
20. Abactinals nonimbricate (0), imbricate (1)
21. Abactinals with simple contact (0), specialized tongue-and-groove articulations (1)
22. Paxillae present (1), absent (0)
23. Numerous spines on each abactinal (0), single spine on each ossicle (1)
24. Abactinal spine without specialized articulation (0), with tubercle-like socket joint (1)
25. Papular openings present only between abactinals (0), between marginals also (1) between actinals also (2)
26. Papulae single (0), clustered into groups (1)
27. Papular areas lack ossicles (0), with small ossicles (1)
28. Epiproctal cone absent (0), present (1)
29. Spines on abactinal ossicles separate (0), webbed (1)
30. Superomarginals absent (0), present (1)
31. Marginals larger than abactinals (0) of similar size (1)
32. Marginals do not form conspicuous border to disc and arms (0), form conspicuous border (1)
33. Marginals not block-like (0), marginals block-like (1)
34. Marginals not paxilliform (0) paxilliform (1)
35. Small ossicles between marginal rows (intermarginals) absent (0), present (1)
36. Intermarginal fascioles absent (0), present (1), modified to form cribriform organs (2)
37. Actinals absent (0), present (1)
38. Actinals irregular (0), set in transverse rows (1)

39. Longest actinal rows parallel to marginals (MRP) (0), parallel to adambulacrals (ARP) (1)

40. Transverse actinal channels/furrows absent (0), present (1)

Pedicellariae

41. Absent (0), present (1)

42. Absent (0), bivalved (1) multivalved (2)

43. Superficial, simply attached to surface (0), third ossicle becomes integral part of ped (1)

44. Third ossicle large primary plate (0), small secondary ossicle (1)

45. Specialized articulation surfaces between third ossicle and valves absent (0), present (1)

46. Longitudinal internal adductor muscles (*ilad*) on third piece absent (0), present (1)

47. Peduncle absent (0), present (1)

48. Transverse adductor (*itad*) between valves lacking (0), present (1)

49. Valves do not articulate at base (0), articulate together at base (1)

50. Internal adductor absent (0), becomes distal adductor on crossed peds (1)

51. Basal piece absent (0), present, forms cup-shaped base to ped (1), integrated as cross-piece (crossed peds) (2)

52. Straight peds only (0), both straight and crossed peds present (1), only crossed (2)

53. Abductor muscles single (0), separate proximal and distal abductors present (crossed peds) (1)

Ambulacral groove

54. Upper transverse ambulacral muscle (*abtam*) absent (0), present (1)

55. Lower transverse ambulacral muscle (*actam*) absent (0), present (1)

56. Facet for upper transverse ambulacral muscle visible on abactinal surface of ambulacral head (0), deeply inset between proximal and distal regions of dentition (1)

57. Dentition comprises single central set of ridges and grooves (0), separate proximal and distal regions (1)

58. Ambulacral base broadest part of ossicle (0), base forms <50 per cent of total breadth (1)

59. Ambulacral head upright with little imbrication (amb angle approx. 90 degrees) (0), moderate proximal imbrication (amb angle 55–70 degrees) (1), strong proximal imbrication (amb angle <45 degrees) (2), upright, vertebra-like (3)

60. Ambulacral head rectangular in abactinal view (0), with triangular proximal extension (1), with distal flange to carry additional interambulacral muscle (2)

61. Ambulacral head not flattened (0), obliquely flattened proximally (1)

62. Longitudinal interambulacral articulation (*lia*) facet small, discrete region of smooth stereom (0), elongated abactinally–actinally (1) large, diffuse (2)

63. Longitudinal interambulacral articulation (*lia*) facet flat (0) concavo-convex (1)

64. Longitudinal interambulacral muscle (*lim*) facets nearly symmetrical (0), strongly asymmetrical with distal facet broad, triangular; proximal facet narrow, deeply incised (1)

65. Ambulacral waist composed of nonlabyrinthic stereom (0), labyrinthic stereom with transversely elongated glassy trabeculae (1)

66. Enlarged wing-like proximal and distal projections for attachment of ambulacral–adambulacral muscles on ambulacral base absent (0), present (1)

67. Processes for amb–adamb muscles on ambs symmetrical (0), strongly asymmetrical (1)

68. Transverse ridges on abactinal surface of ambulacral base absent (0), present (1)

69. Superambulacrals absent (0), present (1)

70. Ambulacrals moderately elongated (0), short, strongly compressed (1)

71. Distal adamb–amb articulation (*ada1*) comprises single transverse facet (0), two discrete facets (*ada1a*, *ada1b*) (1)

72. *Ada1* broad, smooth (0), *ada1* narrow, rugose (1)

73. *Ada2* and *adada* discrete, separate (0) confluent, form concavo-convex structure (1)

74. *Ada3* entirely interadamb articulation with no amb contact (0), has both amb and adamb contact (1)

75. Adambs imbricate distally (0), proximally (1)

76. Adambs twice as broad as ambs (0), as broad or narrower than ambs (1)

77. *Adada* present (0) absent (1)

78. Distal interadamb articulation surfaces comprise two discrete areas of smooth imperforate stereom (*adada*, *ada3*) (0), single (*adada*) (1), no discrete interadamb articulation surfaces (2)

79. Proximal facets *ada2*, *ada3*, separate (0), united to form transverse hourglass-shaped or crescentic facet (1)

80. Adamb lacking extension (0), adamb with elongated abradial extension (1)

81. Abradial muscle between successive extensions absent (0), present (1)

82. Abradial articulation surface between extensions absent (0), present (1)

83. Distal adradial process to adamb absent (0) present (1)

84. Adamb spines in single transverse row (0), discrete groove-parallel furrow and subadambulacral spines (1)

85. Adambulacral spines numerous (0), few (1–3) (1)

Mouth frame

86. Mouth frame in internal abactinal view pentagonal (0), stellate (1), ring-like, flexible (2), ring-like, inflexible (3)

87. Odontophore deeply inset between circumoral ossicles (0), relatively shallow in position, just beneath heads of circumoral ossicles (1), flush with circumoral heads (2)

88. Proximal lateral process of odontophore projects into base of first podial opening (0), does not project into podial opening (1)

89. Lateral notch on odontophore present, fits around distal bar of circumoral (0), notch absent, side of odontophore parallel with distal bar (1)

90. Oral ossicles large, project proximally to occupy much of peristomial ring (>50 per cent) (0), smaller, project slightly (<50 per cent) (1), very small, project little (<10 per cent) (2)

91. Adoral carina absent (0), present (1)

92. Actinostome absent (0), present (1)

93. Odontophore/axillary elongate T-shaped (0), short, broad with transverse bar (1), rhomboidal-trapezoidal (2), block-like, rectangular (3), fusiform, elongated external face (4)

94. Odontophore biconvex (0), abactinally convex, actinally concave (1), actinally convex, abactinally flat (2)

95. Odontophore articulates with marginals (0), does not articulate with marginals (1)

96. Keel of odontophore deep (0), shallow (1)

97. External face on odontophore/axillary present (0), entirely absent (1)

98. Odontophore articulates freely with oral ossicles (0), immovably attached to or fused with oral ossicles (1)

99. Odontophore does not articulate with circumoral ossicle (0), articulates distally with circumoral ossicle (1)

100. Odontophore does not form part of wall of first podial internal opening (0), forms integral part of wall of first opening (1)

101. Keel present on odontophore (0), absent (1)

102. Proximal and distal oral articulation structures on odontophore (*poda*, *doda*) dissimilar in size and shape (0), similar, arranged in rhomboidal pattern (1)

103. Specialized articulation surface for chevron plates on abactinal distal odontophore surface absent (0), present (1)

104. Odontophore-oral muscle insertion extends to actinal margin of oral ossicle (0), restricted to concave odontophore capsule (1)

105. Distal bar of circumorals strongly angled to body of ossicle (0), nearly parallel with body of ossicle (1).

106. Lateral view of proximal circumoral bar shows double articulation with oral (0), does not show double contact (1)

107. Circumoral ossicles higher than long (0), longer than high (1)

108. Proximal oral-circumoral articulation small, narrow (0), flat, broad (1)

109. Circumoral ossicles upright, nonimbricate (0), imbricate distally with proximal ambulacrals (1)

110. Upper transverse intercircumoral muscle facet broad (0), largely internal (1)

111. Circumoral heads short (=1–2 adjacent ambulacrals), elongated (=3–6 ambs) (1)

112. Circumoral heads articulate by means of dentition (0), articulation flat, fixed, immobile (1)

113. Articulation structures between orals of a pair (*iioa*, interradial surface) weakly developed (0), made up of fine dentition (1), discrete flat articulation surfaces, made of imperforate stereom (2)

114. Apophyse of oral low, <25 per cent total height of ossicle (0), tall, more than 25 per cent total height (1)

115. Apophyse vertical (0), inclined proximally at 10–20 degrees (1)

116. Abactinal interradial interoral muscle (*abiim*) attachment shallow, positioned at base of apophyse (0), deep, set on proximal circumoral bar (1)

117. Groove for radial vessel (*rvg*) on inner face of oral apophyse short (0), elongated (1)

118. Oral without proximal blade (0), with triangular proximal blade-like extension (1)

119. Groove for ring nerve (*rng*) deeply embayed (0), shallow, superficial (1)

120. Articulation between oral and first adambulacral oblique, visible on lateral face of oral (0), set at 70–90 degrees to lateral face of oral (1)

121. First adambulacrals relatively small in proportion to orals (0), enlarged, with broad proximal articulation surface (1), extend along length of orals (2)

122. Actinal surface of orals forms largest part of plate (0), actinal surface of orals small, inconspicuous (1)

123. Actinal surface of orals raised, rounded (0), flattened (1)

124. External (radial) surfaces of orals lack articulation with adjacent orals (0) flattened articulation surfaces present (1)

125. External (actinal) part of orals comprises single surface (0), made up of discrete rounded lateral rim and central (sutural) mound (1)

126. Spines on oral ossicles not differentiated into oral and suboral groups (0), discrete oral and suboral spines present (1)

127. Oral spines discrete do not articulate with each other (0), form dense, articulating cover over actinal part of oral ossicles (1)

128. Oral spines numerous (0), few (<3) (1).